D1234811

Gâteaux Differentiability of Convex Functions and Topology

CANADIAN MATHEMATICAL SOCIETY SERIES OF MONOGRAPHS AND ADVANCED TEXTS

Monographies et Études de la Société mathématique du Canada

EDITORS/RÉDACTEURS: Jonathan M. Borwein and Peter B. Borwein

A complete list of titles in this series appears at the end of this volume.
Tous les titres de cette collection sont énumérés à la fin du volume.

Gâteaux Differentiability of Convex Functions and Topology

Weak Asplund Spaces

MARIÁN J. FABIAN

A Wiley-Interscience Publication

JOHN WILEY & SONS, INC.

New York · Chichester · Weinheim · Brisbane · Singapore · Toronto

This text is printed on acid-free paper.

Library of Congress Cataloging in Publication Data:

Fabian, Marian J.
    Gâteaux differentiability of convex functions and topology /
Marian J. Fabian.
        p.   cm. — (Canadian Mathematical Society series of monographs
and advanced texts)
    "A Wiley-Interscience publication."
    Includes bibliographical references and index.
    ISBN 0-471-16822-X (cloth : alk. paper)
    1. Asplund spaces. 2. Convex functions. I. Title. II. Series.
QA322.2.F33   1997
    515′.743—dc20   96-36642

Printed in the United States of America

10 9 8 7 6 5 4 3 2 1

To Pavla

Contents

Preface

Asplund (weak Asplund) spaces are those Banach spaces which have the property that every convex continuous function defined on them is Fréchet (Gâteaux) differentiable at the points of a dense G_δ subset. The class of Asplund spaces, since its definition by Asplund in 1968, has flourished to a beautiful and huge theory. In particular, many quite different equivalent characterizations of Asplund spaces exist. Although there is no special book devoted to this topic, it would not be difficult to arrange such a text, simply by putting together sections and chapters from the books of J. Giles [Gi]; R. R. Phelps [Ph]; J. Diestel [Di1]; J. Diestel and J. J. Uhl, Jr. [DU]; and R. DeVille, G. Godefroy, and V. Zizler [DGZ2].

The situation with the weak Asplund spaces is far from analogous. Currently there are no equivalent conditions for a Banach space to be weak Asplund and only a modest number of necessary conditions are known. Also, only very few permanence properties are known to hold. The main interest in weak Asplund spaces has focused on finding sufficient conditions, that is, in looking for larger and larger well-behaved subclasses of the class of weak Asplund spaces. In this way, several classes of Banach spaces and even topological spaces have appeared and various original techniques have been developed. This brings us to the aim of this text: To put together some less folklore results (and their nice proofs) related more or less to the concept of weak Asplund spaces. Thus we study Asplund spaces, Asplund generated spaces (known in the literature as Grothendieck–Šmulyan generated (GSG) spaces), weakly compactly generated (WCG) spaces, Vašák (that is, weakly countably determined) spaces, and Banach spaces whose dual belong to Stegall's class and the relations between them. All of these classes have more or less to do with topology. We thus work with various classes of topological spaces: Eberlein, Gul'ko, Radon–Nikodým compacta, Stegall's class, fragmentability, and so on.

Chapter 1 starts with Asplund spaces, a quite natural subclass of the weak Asplund spaces. Asplund spaces are quickly reviewed. Then WCG spaces and Eberlein compacta are studied. Further, via dense continuous linear images, the Asplund generated spaces are defined and investigated. Some questions about Asplund sets and Radon–Nikodým compacta are also discussed. Perhaps a culminating result here is that a compact space K is a Radon–Nikodým compact if and only if the Banach space $C(K)$ of continuous functions on K is Asplund generated.

Chapter 2 lists all known (to the author's knowledge) properties of weak Asplund spaces. Of particular interest is the fact that a compact space K contains a dense, completely metrizable subset whenever $C(K)$ is weak Asplund. In this chapter we also provide a counterexample, which shows that the weak Asplundness is not a three-space property.

Chapter 3 deals with Stegall's class of topological spaces as well as with its Banach space counterpart. This is so far the largest (well-behaved) subclass of weak Asplund spaces, and it is not clear if this inclusion is proper. We focus here on permanence properties of Stegall's classes.

Chapter 4 shows that two significant classes belong to Stegall's class of Banach spaces: weakly $\mathcal{K}$-analytic spaces and spaces with Gâteaux smooth norm. These two results can be obtained from more general, respectively stronger theorems given in Chapters 5 and 7. We included them because of the beauty of their proofs. It is also shown here that a weakly $\mathcal{K}$-analytic space may not be a subspace of an Asplund generated space.

Chapter 5 is devoted to fragmentable topological spaces and their Banach space counterparts. These classes are smaller (at least formally) than the corresponding Stegall classes. Results in a flavor similar to Chapter 3 are presented here. It is also shown that Gâteaux smooth Banach spaces have fragmentable dual and hence are weak Asplund.

Chapter 6 has an auxiliary character supporting Chapters 7 and 8. A technique of constructing "long sequences" of linear projections on certain Banach spaces is developed here. We show how strong impact on the space has the presence of such a "long sequence".

In Chapter 7 the theory of $\mathcal{K}$-countably determined spaces, Vašák spaces (that is, weakly countably determined spaces), and Gul'ko compacta is presented. Among other things it is proved that Gul'ko compacta are Corson and that Vašák spaces have fragmentable dual. In an example, it is shown that the weak Asplundness goes beyond Vašák spaces.

A main achievement of Chapter 8 says that a Banach space is (a subspace of) a WCG space if and only if it is both (a subspace of) an Asplund generated space and a Vašák space, or more generally, a weakly Lindelöf determined space. This has a nice topological counterpart: A compact is Eberlein if and only if it is simultaneously a continuous image of a Radon–Nikodým compact and a Gul'ko, or more generally, a continuous image of a Corson compact. The chapter ends by a construction of a weakly $\mathcal{K}$-analytic space not containing ℓ_1 which is not a subspace of an Asplund generated space.

The offered text is mainly addressed to functional analysts with emphasis on nonseparable Banach spaces and differentiability. We believe, however, that it will also attract some topologically oriented readers because half of the material presented here has a topological flavor. Each chapter contains open, often quite untouched questions, so that we hope that the presented text will prove interesting for researchers and in particular for Ph.D. students.

We thank our colleagues and students who provided us with their suggestions and criticism. In particular, we mention Warren B. Moors from Auckland University, who was willing to carefully read the entire text and as a result improved it in many places. We would also like to thank The Mathematical Institute of The Czech Academy of Sciences for their support during the final stage of work on the manuscript.

Gâteaux Differentiability of Convex Functions and Topology

Introduction

When we write $A := B$ or $B =: A$, we understand that the meaning of the letter A is defined by the expression B.

If M is a set, $\#M$ means its cardinality. The first infinite cardinal and ordinal are denoted by $\aleph_0$ and ω_0, respectively. The first uncountable cardinal and ordinal are denoted by $\aleph_1$ and ω_1, respectively. Given an ordinal α, the cardinality of the interval $[0, \alpha)$ is denoted by card α.

The letter IR is reserved for the set of real numbers, provided with its natural topology. The letter IN means the set of positive integers, eventually provided with the discrete topology. If A is a nonempty subset of IR, we put $\sup A = \sup\{a : a \in A\}$ and $\inf A = \inf\{a : a \in A\}$.

If X is a set, then 2^X denotes the family of all subsets of X. Given nonempty sets X, Y, the expression $F : X \to 2^Y$ means that $\emptyset \neq F \subset X \times Y$, and we say that F is a *multivalued mapping* from X to Y. Further, for $x \in X$ we put $Fx = \{y \in Y : (x, y) \in F\}$. Of course, if $\#Fx \leq 1$ for all $x \in X$, then F is the usual (single-valued) mapping from X to Y and we write $F : X \to Y$. For $F : X \to 2^Y$ we put $D(F) = \{x \in X : Fx \neq \emptyset\}$, $R(F) = \{y \in Y : (x, y) \in F$ for some $x \in X\}$ and call these sets the *domain* and the *range* of F, respectively. If $R(F) = Y$, we say that F is *surjective* or *onto*. Unless otherwise stated, we shall assume that $D(F) = X$. Further, we put $F^{-1} = \{(y, x) : (x, y) \in F\}$, so that $F^{-1} : Y \to 2^X$. If $F : X \to 2^Y$ and $Z \subset X$, the symbol $F_{|Z}$ means the restriction of F to Z, that is, $F_{|Z} : Z \to 2^Y$ and $F_{|Z}x = Fx$ for all $x \in Z$. If $F : X \to 2^Y$ and $G : Y \to 2^Z$ are two mappings, then $G \circ F$ denotes their composition, that is, $(G \circ F)(x) = G(F(x))$, $x \in X$; so $G \circ F : X \to 2^Z$. A single-valued mapping $F : X \to Y$ is called *injective* or *one-to-one* if $Fx \neq Fx'$ whenever $x \neq x'$. By a function we understand a single-valued mapping whose range lies in the set of real numbers. For a function $f : X \to$ IR, we put $\operatorname{supp} f = \{x \in X : f(x) \neq 0\}$ and call this set the *support* of f.

Topological spaces are always assumed to be Hausdorff and usually nonempty. Let X be a topological space. If $A \subset X$, then $\bar{A}$ or cl A denote the closure of the set A while int A means the interior of A. If there are several topologies on a set X and we want to stress that the closure is with respect to a topology τ, we write $\overline{A}^{\tau}$. A *density* of a set $A \subset X$, denoted by dens A, is defined as the smallest cardinal of the form $\#D$, where $D \subset A$ and $\overline{D} \supset A$. If X_{γ}, $\gamma \in \Gamma$, are topological spaces, we consider $\prod_{\gamma \in \Gamma} X_{\gamma}$ with the product topology. This topology is called *pointwise* topology and is denoted by p. A set in X is called G_{δ} if it can be expressed as the intersection of a countable family of open sets in X. If such a set is a singleton, we speak about a G_{δ} point. A set in X is called F_{σ} if its complement is G_{δ}. A subset M of X is called *residual* if there are open dense sets $G_n \subset X$, $n = 1, 2, \ldots$, such that $M \supset \bigcap_{n=1}^{\infty} G_n$, or equivalently, if $X \backslash M$ is of the first (Baire) category. Clearly, a countable intersection of residual sets is also residual. A topological space is called a *Baire* space if all its residual subsets are dense in it. Complete metric spaces, in particular Banach spaces, are Baire spaces [En, Theorems 3.9.3, 4.3.26]. A topological space X is called *completely regular* if for every closed set $C \subset X$ and every $x \in X \backslash C$ there is a continuous function $f : X \to [0,1]$ such that $f(x) = 0$ and $f(c) = 1$ for all $c \in C$. We recall that *X is completely regular if and only if there is a compact space K such that X is a subspace of K* [En, Theorem 3.2.6].

Banach spaces in this text are always considered over the field of real numbers. If we work in Banach spaces with topological concepts like continuity, closure, and so on, and no concrete topology is specified, we always assume the topology generated by the norm. Let V be a Banach space. If $A, B \subset V$, we define $A + B = \{a + b : a \in A, b \in B\}$. Unless otherwise specified, we assume that the norm on V is denoted by $\|\cdot\|$. By a *subspace* of V we understand any closed linear subset of it. If Y is a subspace of V, then the symbol V/Y denotes the quotient, that is, the Banach space consisting of elements $[v] := v + Y$, with the norm $\|[v]\| := \inf\{\|v + y\| : y \in Y\}$. If M is a subset of V, then sp$M$ means the set

$$\left\{ \sum_{i=1}^{n} \lambda_i v_i : \lambda_i \in \mathbb{R}, \ v_i \in V, \ i = 1, \ldots, n, \ n \in \mathbb{N} \right\};$$

sometimes we write $\overline{\text{sp}}M$ instead of $\overline{\text{sp}M}$. The convex hull co$M$ of $M \subset V$ is defined by

$$\text{co}M = \left\{ \sum_{i=1}^{n} \lambda_i v_i : \lambda_i \geq 0, \ v_i \in V, \ i = 1, \ldots, n, \ \sum_{i=1}^{n} \lambda_i = 1, \ n \in \mathbb{N} \right\};$$

thus M is convex if and only if co$M = M$. The closed unit ball of V is denoted by B_V, that is, $B_V = \{v \in V : \|v\| \leq 1\}$. The symbol V^* means the topological dual of V, that is, the Banach space of all linear continuous mappings $v^* : V \to \mathbb{R}$ equipped by the norm $\|v^*\| = \sup\{\langle v^*, v \rangle : v \in B_V\}$, so that we use the same symbol for the norm and the corresponding dual norm. For $v^* \in V^*$ and $v \in V$, we put $\langle v^*, v \rangle = v^*(v)$. The second dual of V, that is, the dual of V^*, is denoted by V^{**}. Then there exists a canonical linear isometry $\kappa : V \to V^{**}$ defined as

$$\langle \kappa(v), v^* \rangle = \langle v^*, v \rangle, \qquad v \in V, \qquad v^* \in V^*.$$

We always think of V as a subspace of V^{**}. The letter w is reserved for denoting the weak topology while w^* means the weak* topology. If $M \subset V^*$, then $\overline{M}^{w^*}$ means the closure of M in the weak* topology. Let $T : V \to Y$ be a linear mapping between Banach spaces V and Y. We put $\|T\| = \sup\{\|Tv\| : v \in B_V\}$. Of course, $\|T\| < \infty$ if and only if T is continuous. We define an adjoint mapping $T^* : Y^* \to V^*$ of T by $\langle T^*y^*, v \rangle = \langle y^*, Tv \rangle$, $v \in V$, $y^* \in Y^*$; note that $\|T\| = \|T^*\|$.

Let $\| \cdot \|$ be a norm on a Banach space V. We say that $\| \cdot \|$ is *strictly convex* if $v_1 = v_2$ whenever $v_1, v_2 \in V$ and $\frac{1}{2}\|v_1 + v_2\| = \|v_1\| = \|v_2\|$. The norm $\| \cdot \|$ is called *locally uniformly rotund (LUR)* if $\|v_n - v\| \to 0$ whenever $v, v_n \in V$ and

$$2\|v\|^2 + 2\|v_n\|^2 - \|v + v_n\|^2 \to 0.$$

Let Γ be a nonempty set. We use the symbol $\ell_\infty(\Gamma)$ to denote the (linear) space of all $x = \{x_\gamma : \gamma \in \Gamma\} \in \mathrm{IR}^\Gamma$ such that

$$\|x\|_\infty := \sup\{|x_\gamma| : \gamma \in \Gamma\} < \infty.$$

We note that $(\ell_\infty(\Gamma), \| \cdot \|_\infty)$ is a Banach space. The symbol $c_0(\Gamma)$ means its subspace consisting from $x = \{x_\gamma\} \in \ell_\infty(\Gamma)$ such that the set $\{\gamma \in \Gamma : |x_\gamma| > \epsilon\}$ is finite for all $\epsilon > 0$. Given $p \in [1, +\infty)$, the symbol $\ell_p(\Gamma)$ denotes the space

$$\left\{ x = \{x_\gamma : \gamma \in \Gamma\} \in \mathrm{IR}^\Gamma : \|x\|_p := \left(\sum_{\gamma \in \Gamma} |x_\gamma|^p \right)^{1/p} < \infty \right\}.$$

Here $(\ell_p(\Gamma), \| \cdot \|_p)$ is also a Banach space. If $\Gamma = \mathrm{IN}$, we simply write ℓ_∞, c_0, ℓ_p. If K is a compact space, then the (Banach) space of continuous functions on it endowed with the supremum norm $\| \cdot \|$ is denoted by $C(K)$. The letter p then means the topology on $C(K)$ inherited from the product topology of IR^K.

Given a convex set $C \subset V$ containing 0, its Minkowski functional is defined as

$$v \mapsto \inf\{\lambda \geq 0 : v \in \lambda C\}, \qquad v \in V.$$

Consider a function $f : V \to \mathrm{IR}$ and a point $v \in V$. We say that f is *Gâteaux differentiable* at v if there exists $\xi \in V^*$ such that

$$\frac{1}{t}\left[f(v + th) - f(v) - \langle \xi, tv \rangle \right] \to 0 \quad \text{as } t \to 0$$

for every $h \in V$. Then ξ is called a *Gâteaux derivative* of f at v. If moreover the above limit is uniform with respect to all $h \in B_V$, we speak about *Fréchet differentiability* and *Fréchet derivative* of f at v. We say that the function f is *Gâteaux (Fréchet) smooth* if it is Gâteaux (Fréchet) differentiable at each point $v \in V$. Of course, if f is a norm, we exclude the point $v = 0$. The function f is called *convex* if

$$f(\alpha u + \beta v) \le \alpha f(u) + \beta f(v)$$

for all $u, v \in V$ and all $\alpha, \beta \ge 0$, with $\alpha + \beta = 1$. It is easy to check that a convex function $f: V \to \mathrm{IR}$ is Gâteaux (Fréchet) differentiable at $v \in V$ if and only if

$$\frac{1}{t} \left[f(v + th) + f(v - th) - 2f(v) \right] \to 0 \quad \text{as} \quad t \downarrow 0 \quad \text{for all} \quad h \in V$$

$$\left(\frac{1}{\|h\|} \left[f(v + h) + f(v - h) - 2f(v) \right] \to 0 \quad \text{as} \quad h \in V \text{ and } h \to 0 \right).$$

For a convex function $f: V \to \mathrm{IR}$ and $v \in V$ we define

$$\partial f(v) = \left\{ v^* \in V^* : f(v + h) - f(v) \ge \langle v^*, h \rangle \text{ for all } h \in V \right\};$$

This set is called the *subdifferential* of f at v. It is easy to check that $\partial f(v)$ is a convex weak* compact set. If f is moreover continuous at v, then $\partial f(v)$ is nonempty. In this way, we have defined a multivalued mapping $\partial f: V \to 2^{V^*}$. The convexity of f implies that ∂f is monotone, that is,

$$\langle v_1^* - v_2^*, v_1 - v_2 \rangle \ge 0 \quad \text{whenever} \quad v_i \in V, \quad v_i^* \in \partial f(v_i), \quad i = 1, 2;$$

see [Ph, Example 2.2(a)]. Another important and easily checked feature of the mapping ∂f is its norm-to-weak* upper semicontinuity, which means that the set $\{v \in V : \partial f(v) \cap C \ne \emptyset\}$ is closed whenever C is a weak* closed subset of V^* [Ph, Proposition 2.5]. Note also that a convex continuous function f is Gâteaux differentiable at v if and only if $\partial f(v)$ is a singleton.

Throughout the text we use several standard theorems from functional analysis and topology. Let us mention those facts which are referred to most frequently: Zorn's lemma [DS, Theorem I.2.7], Baire's category theorem [En, Theorem 3.9.3], the Hahn–Banach theorem [DS, Theorem II.3.11] and its geometrical counterpart, the separation theorems [DS, Theorem V.2.8, Corollary V.2.12], Banach's open mapping theorem [DS, Theorem II.2.1], the Banach–Steinhaus theorem [DS, Theorem II.1.17], Goldstine's theorem [DS, Theorem V.4.5], Alaoglu's theorem [DS, Theorem V.4.2], the Stone–Weierstrass theorem [DS, Theorem IV.6.16], F. Riesz's representation theorem [DS, Theorem IV.6.3], Lebesgue's dominated convergence theorem [DS, Corollary III.6.16], the Eberlein–Šmulyan theorem [DS, Theorem V.6.1], the

Krein–Šmulyan theorem [DS, Theorem V.6.4], the Tietze–Urysohn theorem [En, Theorem 2.1.8]. Besides [DS, En], we recommend the books [Ku, KN, Ru1, Ru2, RR] or any other textbook from functional analysis and topology. More information concerning convex functions may be found in [Ph, Gi]. For the geometry of Banach spaces we refer to [DGZ2, Di1, Bo, DU, HHZ].

Chapter One

Canonical Examples of Weak Asplund Spaces

Definition 1.0.1. A Banach space V is called *Asplund (weak Asplund)* if every continuous convex function on it is Fréchet (Gâteaux) differentiable at each point of a dense G_δ subset of V.

Clearly, both these concepts are isomorphically invariant. Because, trivially, Fréchet differentiability implies Gâteaux differentiability, we have:

Proposition 1.0.2.

Every Asplund space is weak Asplund.

1.1. ASPLUND SPACES

There exist many equivalent characterizations of the Asplund spaces. Below we list those which will be of use later in this and subsequent chapters. For a nonempty bounded set A in the dual V^* of a Banach space V, for $e \in V$, and for $\alpha > 0$, we put

$$S(A, e, \alpha) = \left\{ v^* \in A : \langle v^*, e \rangle > \sup \langle A, e \rangle - \alpha \right\}$$

and call it a *weak* slice* of A.

Theorem 1.1.1.

For a Banach space V the following assertions are equivalent:
  (i) V is Asplund;
  (ii) Y^* is separable whenever Y is a separable subspace of V; and

(iii) V^* *is weak* dentable, that is, for every nonempty bounded set $A \subset V^*$ and for every $\epsilon > 0$ there exist $0 \neq e \in V$ and $\alpha > 0$ such that the weak* slice $S(A, e, \alpha)$ has norm diameter less than ϵ.*

For proofs we refer to [DGZ2, Section I.5], [Ph, Chapter 2], or eventually [As, NP, Gi]. Thus all reflexive spaces, and in particular ℓ_p and L_p, with $1 < p < +\infty$, are Asplund spaces; see [DS, Corollary IV.8.2]. Further, $c_0(\Gamma)$ with any set Γ, $C([0, \mu])$, for any ordinal μ (see Theorem 1.1.3), James's and long James's space [Bo, Section 7.7], the space of Johnson and Lindenstrauss [JL1, JL2] and many other spaces are Asplund. On the other hand, ℓ_1, ℓ_∞, L_1, L_∞, $C([0, 1])$, and many other spaces are not Asplund.

Theorem 1.1.1 will be formulated (and proved) in a more general setting in Theorem 1.4.5.

The next theorem shows that the Asplund spaces enjoy several permanence properties, thus enabling us to construct new Asplund spaces from old ones.

Theorem 1.1.2

(i) *If V is an Asplund space, Y is a Banach space, and $T : Y \to V$ is a continuous linear mapping with T^*V^* dense in Y^*, then Y is Asplund; in particular, every subspace of V is an Asplund space.*

(ii) *If V is Asplund and $T : V \to Y$ is linear, continuous, and surjective, then Y is Asplund; in particular, quotients of an Asplund space are Asplund.*

(iii) *If Y is a subspace of a Banach space V and both Y and the quotient V/Y are Asplund, then so is V.*

(iv) *If Γ is any nonempty set and for every $\gamma \in \Gamma$, $(V_\gamma, \|\cdot\|_\gamma)$ is an Asplund space, then the spaces $\left(\sum_{\gamma \in \Gamma} V_\gamma\right)_{c_0}$ and $\left(\sum_{\gamma \in \Gamma} V_\gamma\right)_{l_p}$, with $1 < p < +\infty$, are Asplund.*

(v) *If V is Asplund and (Ω, Σ, μ) is a finite measure space, then $L_p(\Omega, \Sigma, \mu, V)$, with $1 < p < +\infty$, is an Asplund space.*

Proof. Let V, Y, and T be as in (i). Consider any separable subspace Z of Y. According to Theorem 1.1.1, it is enough to show that Z^* is separable. Assume this is not the case. Then there are $\epsilon > 0$ and an uncountable set $\{\zeta_\alpha : \alpha \in A\}$ in Z^* such that $\|\zeta_\alpha - \zeta_\beta\| > \epsilon$ whenever $\alpha, \beta \in A$ and $\alpha \neq \beta$. For $\alpha \in A$ we find $\eta_\alpha \in Y^*$ such that $\eta_{\alpha|Z} = \zeta_\alpha$. Now for each $\alpha \in A$ we take $\xi_\alpha \in V^*$ such that $\|T^*\xi_\alpha - \eta_\alpha\| < \epsilon/3$. Then

$$\sup\langle \xi_\alpha - \xi_\beta, TB_Z \rangle = \sup\langle T^*\xi_\alpha - T^*\xi_\beta, B_Z \rangle$$
$$> \sup\langle \eta_\alpha - \eta_\beta, B_Z \rangle - 2\epsilon/3 = \|\zeta_\alpha - \zeta_\beta\| - 2\epsilon/3 > \epsilon/3$$

whenever $\alpha \neq \beta$. But T is continuous, so TB_Z is a bounded set. Denote, now, $Y_1 = \overline{TZ}$; this is a separable subspace of Y since Z was separable. The above estimate shows that the set $\{\xi_{\alpha|Y_1} : \alpha \in A\}$ is discrete and uncountable. But this contradicts the fact that, according to Theorem 1.1.1, Y_1^* is separable.

Let $T: V \to Y$ be as in (ii). Since T is surjective, by Banach's open mapping theorem, T is an open mapping. Consequently, T^* is an isomorphism of Y^* into V^*. Therefore there is $c > 0$ such that $\|T^* y^*\| \geq c\|y^*\|$ for all $y^* \in Y^*$. Let A be a bounded subset of Y^* and choose $\epsilon > 0$. Then $T^*(A)$ is a bounded subset of V^* and, by Theorem 1.1.1, there is $0 \neq e \in V$ and $\alpha > 0$ such that the weak* slice $S(T^*(A), e, \alpha)$ has diameter less then $c\epsilon$. But we can see that $S(T^*(A), e, \alpha) = T^*(S(A, Te, \alpha))$. Therefore, the weak* slice $S(A, Te, \alpha)$ has diameter less than $c\epsilon/c = \epsilon$ and hence Theorem 1.1.1 guarantees that Y is Asplund.

Let V and Y be as in (iii). Denote by Q the mapping assigning the class $v + Y$ to $v \in V$. Take any separable subspace Z in V. According to Theorem 1.1.1, it is enough to show that Z^* is separable. Since Z is separable, there is a countable set C_1 in Y such that $\|Qz\| = \inf \|z + C_1\|$ for each $z \in Z$. Denote $V_1 = \overline{\mathrm{sp}}(Z \cup C_1)$; this will be again a separable subspace of V. Now we find a countable set C_2 in Y such that $\|Qz\| = \inf \|z + C_2\|$ for each $z \in V_1$ and $C_2 \supset C_1$. Next put $V_2 = \overline{\mathrm{sp}}(V_1 \cup C_2)$. Going on in an obvious way, we shall subsequently construct separable subspaces $(Z \subset) V_1 \subset V_2 \subset \cdots \subset V$ and countable sets $C_1 \subset C_2 \subset \cdots \subset Y$ such that for every $n = 1, 2 \ldots$ we have $\|Qz\| = \inf \|z + C_{n+1}\|$ for each $z \in V_n$. Put $V_0 = \overline{\bigcup V_n}$ and $Y_0 = \overline{\mathrm{sp} \bigcup C_n}$. Then V_0 will be a separable (linear) subspace of V and Y_0 will be a separable (linear) subspace of Y. Moreover, it is easy to check that the mapping $v + Y_0 \mapsto v + Y$ maps V_0/Y_0 linearly and isometrically into V/Y. But we know that V/Y is Asplund. So, by (i), V_0/Y_0 is also Asplund.

Next we realize that Y_0^* is isometric with $V_0^*/Y_0^\perp$ and $(V_0/Y_0)^*$ with $Y_0^\perp$. Thus $V_0^*/Y_0^\perp$ and $Y_0^\perp$ are separable since Y_0 and V_0/Y_0 were both separable Asplund spaces. Now it is an exercise to show that V_0^* must also be separable (because separability is a three-space property). Finally, Z^* is separable since it is a continuous image of the separable space V_0^*.

(iv) We recall that

$$\left(\sum_{\gamma \in \Gamma} V_\gamma \right)_{c_0} = \left\{ v = \{v_\gamma : \gamma \in \Gamma\} \in \prod_{\gamma \in \Gamma} V_\gamma : \{\|v_\gamma\|_\gamma : \gamma \in \Gamma\} \in c_0(\Gamma) \right\},$$

and the norm $\| \cdot \|$ on this space is defined by $\|\{v_\gamma\}\| = \max\{\|v_\gamma\|_\gamma : \gamma \in \Gamma\}$. The space $\left(\sum_{\gamma \in \Gamma} V_\gamma \right)_{\ell_p}$ is defined similarly. Let Y be any separable subspace of $\left(\sum_{\gamma \in \Gamma} V_\gamma \right)_{c_0}$, where all V_γ are Asplund. Then Y contains a dense sequence $\{y_n\}$. For each $n \in \mathrm{IN}$ we write $y_n = \{v_\gamma^n : \gamma \in \Gamma\}$. Then there exist countable sets $\Gamma_n \subset \Gamma$ such that $v_\gamma^n = 0$ whenever $\gamma \in \Gamma \backslash \Gamma_n$. Denote $\Gamma_0 = \bigcup \Gamma_n$; this set will also be countable. For $\gamma \in \Gamma_0$ we put $Z_\gamma = \overline{\mathrm{sp}}\{v_\gamma^n : n \in \mathrm{IN}\}$; thus Z_γ are separable. Then clearly Y is isometric with a subspace of $Z := \left(\sum_{\gamma \in \Gamma_0} Z_\gamma \right)_{c_0}$. Now, we can easily see that Z^* is isometric with $\left(\sum_{\gamma \in \Gamma_0} Z_\gamma^* \right)_{\ell_1}$ (see [Tay, Exercise 4.32.2]), and that this space is separable since each Z_γ^* is separable by Theorem 1.1.1. Hence Y^* is separable as well. Thus Theorem 1.1.1 ensures that $\left(\sum_{\gamma \in \Gamma} V_\gamma \right)_{c_0}$ is Asplund. That $\left(\sum_{\gamma \in \Gamma} V_\gamma \right)_{\ell_p}$ is Asplund can be shown in a similar way or using (i) and what we have just proved.

(v) We recall that $L_p(\Omega, \Sigma, \mu, V)$ is the (Banach) space of all (equivalence classes of) Bochner integrable functions f from Ω into V such that the norm

$$\|f\| := \left(\int_\Omega \|f(\omega)\|^p \, d\mu(\omega) \right)^{1/p}$$

is finite. Let Y be any separable subspace of $L_p(\Omega, \Sigma, \mu, V)$. We are to show that Y^* is separable. Then there exists a sequence $\{f_n\} \subset L_p(\Omega, \Sigma, \mu, V)$ such that each f_n assumes finitely many values only and $Y \subset \overline{\{f_n\}}$ [Di1, Theorem 6(b), p. 202]. Let A be the set of all the values of f_n, $n = 1, 2, \ldots$. This is a countable set; put $V_1 = \overline{\mathrm{sp}}A$. Also, let Σ_1 be the σ-algebra generated by $\{f_n\}$; this is a countable family. Then Y is isometric with a subspace of $L_p(\Omega, \Sigma_1, \mu, V_1)$. Now V_1 being Asplund [DU, Theorem 6, p. 195] guarantees that $\left(L_p(\Omega, \Sigma_1, \mu, V_1)\right)^*$ is isometric to $L_q(\Omega, \Sigma_1, \mu, V_1^*)$, where $p^{-1} + q^{-1} = 1$. But V_1^* is separable. Thus $L_q(\Omega, \Sigma_1, \mu, V_1^*)$ is separable, and hence so is Y^*. $\square$

Theorem 1.1.3.

For a compact space K the space $C(K)$ is Asplund if and only if K is scattered, that is, every subset of K has an isolated point.

Proof. Assume first that $C(K)$ is Asplund. We observe that the assignment $k \mapsto \delta_k$ (where $\delta_k(k) = 1$ and $\delta_k(k') = 0$ if $k' \in K \backslash \{k\}$) maps K into $(B_{C(K)^*}, w^*)$ homeomorphically. In what follows we shall identify k and δ_k. Let A be a nonempty subset of K. Since, by Theorem 1.1.1, $C(K)^*$ is weak* dentable, there are $f \in C(K)$ and $\alpha > 0$ such that the weak* slice $S(A, f, \alpha)$ $(= \{k \in A : f(k) > \sup f(A) - \alpha\})$ has diameter less than 2. Note that $S(A, f, \alpha)$ is an open nonempty set in A. We shall show that it is a singleton. Assume, by contradiction, there are $k, k' \in S(A, f, \alpha)$, $k' \neq k$. Then $\|k - k'\| < 2$. However, $\|k - k'\| = \sup\{g(k) - g(k') : g \in B_{C(K)}\} = 2$ [En, Theorems 3.1.9, 1.5.10], a contradiction. It means K is scattered.

Conversely, assume that K is scattered and let Y be a separable subspace of $C(K)$. We shall show that Y^* is separable, thus proving that $C(K)$ is an Asplund space. Let $\{f_n\}$ be a sequence dense in B_Y. We define a mapping $\phi : K \to [-1, 1]^{\mathbb{N}}$ as

$$\phi(k)(n) = f_n(k), \qquad n \in \mathbb{N}, \qquad k \in K.$$

Clearly ϕ is continuous. We put $L = \phi(K)$; then L is a metrizable compact. Let $\psi : Y \to \mathbb{R}^L$ be the mapping defined by

$$\psi(y)(l) = y(k), \qquad y \in Y, \qquad l \in L \quad \text{and} \quad k \in \phi^{-1}(l).$$

Here, ψ is well defined for if $\phi(k) = \phi(k')$, then $f_n(k) = f_n(k')$ for all $n \in \mathbb{N}$ and hence $y(k) = y(k')$ as $y \in \overline{\mathrm{sp}}\{f_n\}$.

Fix any $y \in Y$. We shall show that $\psi(y) \in C(L)$. It is elementary to check that $\psi(y)^{-1}(M) = \phi(K \cap y^{-1}(M))$ for every set $M \subset \mathbb{R}$. Now the continuity of y and ϕ and the compactness of K guarantee that $\phi(K \cap y^{-1}(M))$ is a closed set whenever M is closed. Therefore $\psi(y) \in C(L)$. We can also see that ψ is a linear isometry. Therefore, once we show that $C(L)$ is an Asplund space, we are done.

Next, let us focus on L. Put $L^{(0)} = L$. Let α be an ordinal and assume that $L^{(\beta)}$ has already been defined for all $\beta < \alpha$. If α is a limit ordinal, put simply $L^{(\alpha)} = \bigcap_{\beta < \alpha} L^{(\beta)}$. Otherwise let $L^{(\alpha)}$ be the derived set, that is, the set of all accumulation points of $L^{(\alpha-1)}$. The symbol $K^{(\alpha)}$ is defined in a similar way. Now, the continuity of ψ and an easy transfinite induction argument reveal that $L^{\alpha} \subset \psi(K^{(\alpha)})$ for all α. Since K is scattered, $K^{(\beta)} = \emptyset$ for some ordinal β. Then also $L^{(\beta)} = \emptyset$ and $\{L^{(\alpha)}\}$ is a strictly decreasing "long sequence" until $L^{(\gamma)} = \emptyset$ for some ordinal γ. Therefore L is a scattered space.

Let $\mathcal{B}$ be a countable basis for the topology on L; it exists since L is a metrizable compact. Put $\mathcal{B}_0 = \{U \in \mathcal{B} : L \cap U$ is at most countable$\}$ and $\tilde{L} + L \backslash (\bigcup \mathcal{B}_0)$. Assume that $\tilde{L} \neq \emptyset$. Then $\tilde{L}$ has an isolated point, l, say. We find $U \in \mathcal{B}$ so that $\tilde{L} \cap U = \{l\}$. Then $U \notin \mathcal{B}_0$ and hence the set $L \cap U$ is uncountable. Thus $\tilde{L} \cap U$ is also uncountable, which is impossible. Therefore $\tilde{L} = \emptyset$ and so L is an at most countable set. Knowing this, F. Riesz's theorem says that $C(L)^*$ is isometric with ℓ_1. And this space is separable. Therefore Y^* is also separable. $\square$

1.2. WEAKLY COMPACTLY GENERATED SPACES AND EBERLEIN COMPACTA

Definition 1.2.1. A Banach space V is called *weakly compactly generated* *(WCG)* if there is a weakly compact set K in V such that its linear span sp$K$ is dense in V. A compact space is called an *Eberlein* compact if it is homeomorphic to a weakly compact subset of a Banach space.

Canonical examples of WCG spaces are reflexive spaces [DS, Theorem V.4.7]. We observe that separable Banach spaces are also WCG. In fact, let V be separable; thus there is a sequence $\{v_n\}$ in B_V with sp$\{v_n\}$ dense in V. Then it is enough to put $K = \{(1/n)v_n\} \cup \{0\}$. (This set is even norm-compact.) Another example of a WCG space is $c_0(\Gamma)$ with an arbitrarily large set Γ because we can take $K = \{e_\gamma : \gamma \in \Gamma\} \cup \{0\}$, where e_γ are the canonical vectors. For further examples see Theorem 1.2.4.

In what follows we shall present an important interpolation technique. Let V be a Banach space with a sequence $\| \cdot \|_n$, $n = 1, 2, \ldots$, of equivalent norms on it. Consider the space $Z = \left(\sum_{n=1}^{\infty} (V, \| \cdot \|_n) \right)_{\ell_2}$, that is,

$$Z = \left\{ (v_1, v_2, \ldots) \in V^{\mathbb{N}} : \sum_{n=1}^{\infty} \| v_n \|_n^2 < +\infty \right\},$$

and the norm $\||\cdot\||$ on Z is defined as

$$\||(v_1, v_2, \ldots)\|| = \left(\sum_{n=1}^{\infty} \| v_n \|_n^2 \right)^{1/2}, \qquad (v_1, v_2, \ldots) \in Z.$$

Clearly Z is a Banach space. Consider further its subspace $Y = \{(v_1, v_2, \ldots) \in Z : v_1 = v_2 = \cdots\}$ and define the mapping $T : Y \to V$ by $T((v, v, \ldots)) = v$, $(v, v, \ldots) \in Y$.

Lemma 1.2.2.

Under the above assumptions and notations the mapping T is linear, injective, and continuous and T^*V^* is dense in Y^*, that is, T^{**} is injective.

Proof. While the linearity of T is obvious, the continuity of T follows immediately from the fact that $\| \cdot \|_1$ is an equivalent norm on V. As regards the density of T^*V^*, we fix an arbitrary $y^* \in Y^*$. By the Hahn–Banach theorem we find $z^* \in Z^*$ such that $z^*_{|Y} = y^*$. From the definition of Z, we can easily get $\xi_1, \xi_2, \ldots$ in V^* such that $\|\xi_1\|_1^2 + \|\xi_2\|_2^2 + \cdots < +\infty$ (Here we use the same symbols for a norm and its dual norm.) and

$$\langle z^*, (v_1, v_2, \ldots) \rangle = \sum_{n=1}^{\infty} \langle \xi_n, v_n \rangle \quad \text{for all} \quad (v_1, v_2, \ldots) \in Z.$$

Thus for every $y = (v, v, \ldots) \in B_Y$ we have

$$\langle y^*, y \rangle = \langle z^*, y \rangle = \langle z^*, (v, v, \ldots) \rangle = \sum_{n=1}^{\infty} \langle \xi_n, v \rangle,$$

and so for all $m = 1, 2, \ldots$

$$\left\langle y^* - T^* \left(\sum_{n=1}^{m} \xi_n \right), y \right\rangle = \langle y^*, y \rangle - \sum_{n=1}^{m} \langle \xi_n, v \rangle = \sum_{m+1}^{\infty} \langle \xi_n, v \rangle$$
$$\leq \sum_{m+1}^{\infty} \| \xi_n \|_n \| v \|_n \leq \left(\sum_{m+1}^{\infty} \| \xi_n \|_n^2 \right)^{1/2} \left(\sum_{m+1}^{\infty} \| v \|_n^2 \right)^{1/2}.$$

It follows that

$$\left\|\left\| y^* - T^*\left(\sum_{n=1}^{m}\xi_n\right)\right\|\right\| \leq \left(\sum_{m+1}^{\infty}\|\xi_n\|_n^2\right)^{1/2} \to 0 \quad \text{as} \quad m \to \infty.$$

But $T^*\left(\sum_{n=1}^{m}\xi_n\right)$ belongs to T^*V^*. Therefore $\overline{T^*V^*} = Y^*$. $\square$

Now we are ready to prove a famous interpolation theorem. It says that WCG spaces always have something to do with reflexive spaces.

Theorem 1.2.3.

A Banach space V is WCG if and only if there exists a reflexive space Y and a linear, injective, continuous mapping $T : Y \to V$ such that TY is dense in V and T^*V^* is dense in Y^*; in this case (B_{V^*}, w^*) is an Eberlein compact. Moreover, if $K \subset V$ is a weakly compact set such that $\overline{\mathrm{sp}}K = V$, then Y and T can be found in such a way that $K \subset T(B_Y)$.

Proof. As regards to sufficiency, let Y and T be as in the theorem. Then B_Y is weakly compact and so $T(B_Y)$ is a weakly compact set in V. Moreover $\mathrm{sp}\,T(B_Y) = TY$ and we know that $\overline{TY} = V$. Therefore V is WCG.

Conversely, let K be a weakly compact set in V and assume that it is linearly dense in V. For $n = 1, 2, \ldots$ let $\|\cdot\|_n$ be Minkowski's functional of the set $2^n\mathrm{co}(K \cup -K) + 2^{-n}B_V$. Obviously, each $\|\cdot\|_n$ is an equivalent norm on V. Let us consider the spaces Z, Y and the mapping $T : Y \to V$ constructed, prior Lemma 1.2.2, for our V and our $\|\cdot\|_n$. Then for $v \in K$ and for all $n \in \mathbb{N}$ we have $\|2^n v\|_n \leq 1$; so $\||(v, v, \ldots)|\| \leq \left(\sum_{n=1}^{\infty}2^{-2n}\right)^{1/2} < 1$ and hence $(v, v, \ldots) \in B_Y$. Therefore $K \subset T(B_Y)$ and

$$(V =) \ \overline{\mathrm{sp}}K \subset \overline{TY} \subset V.$$

Recalling that we assume $Y \subset Y^{**}$ and $V \subset V^{**}$, we have

$$T^{**}(B_{Y^{**}}) = T^{**}\left(\overline{B_Y}^*\right) \subset \overline{T^{**}(B_Y)}^* = \overline{T(B_Y)}^*$$

$$\subset \overline{\bigcap_{n=1}^{\infty}\left(2^n\mathrm{co}(K \cup -K) + 2^{-n}B_V\right)}^* \subset \bigcap_{n=1}^{\infty}\left(2^n\overline{\mathrm{co}(K \cup -K)}^* + 2^{-n}B_{V^{**}}\right)$$

$$= \bigcap_{n=1}^{\infty}\left(2^n\overline{\mathrm{co}(K \cup -K)} + 2^{-n}B_{V^{**}}\right) \subset \bigcap_{n=1}^{\infty}\left(V + 2^{-n}B_{V^{**}}\right) = V,$$

so

$$T^{**}(B_{Y^{**}}) \subset V.$$

We used here the Goldstine theorem and the Krein–Šmulyan theorem.

From the inclusion obtained above we shall deduce that $B_{Y^{**}} = B_Y$, and hence the reflexivity of Y will be proved. We fix any $y^{**} \in B_{Y^{**}}$ and put $v = T^{**}y^{**}$; the above inclusion guarantees that v belongs to V. By Goldstine's theorem (and according to the definition of Y), there is a net $\{(v_\alpha, v_\alpha, \ldots)\}$ in B_Y, which converges to y^{**} in the weak* topology. Let $v^* \in V^*$ be arbitrary. Then

$$\langle v^*, v \rangle = \langle T^{**}y^{**}, v^* \rangle = \langle y^{**}, T^*v^* \rangle$$
$$= \lim_\alpha \langle T^*v^*, (v_\alpha, v_\alpha, \ldots) \rangle = \lim_\alpha \langle v^*, T(v_\alpha, v_\alpha, \ldots) \rangle = \lim_\alpha \langle v^*, v_\alpha \rangle.$$

This holds for every $v^* \in V^*$, so $v_\alpha \to v$ weakly. Let us observe that for $m = 1, 2, \ldots$ we have

$$\sum_{n=1}^m \|v\|_n^2 \le \liminf_\alpha \sum_{n=1}^m \|v_\alpha\|_n^2 \le \liminf_\alpha \||(v_\alpha, v_\alpha, \ldots)\||^2 \le 1.$$

Hence $y := (v, v, \ldots)$ belongs to Y, even to B_Y, and we may write $v = Ty$. Summarizing we got that

$$T^{**}y^{**} = v = Ty \ (= T^{**}y).$$

Now, Lemma 1.2.2 says that T^{**} is injective. Therefore $y^{**} = y$ and so $y^{**} \in B_Y$.

It remains to show that (B_{V^*}, w^*) is an Eberlein compact. We already know that $\overline{TY} = V$. It means that T^* is injective. T^* is also weak*-to-weak* continuous. Hence (B_{V^*}, w^*) is homeomorphic with the weak* compact set $T^*(B_{V^*}) \subset Y^*$ which is weakly compact since Y is reflexive.   □

Theorem 1.2.4.

For a compact space K the following assertions are equivalent:
  (i) K is an Eberlein compact;
  (ii) K can be found, up to a homeomorphism, in $(c_0(\Gamma), w)$ for some set Γ;
  (iii) $C(K)$ is WCG;
  (iv) $(B_{C(K)^*}, w^*)$ is an Eberlein compact;
  (v) There exists a family $\mathcal{U} = \bigcup_{n=1}^\infty \mathcal{U}_n$ of open F_σ sets in K such that $\mathcal{U}$ is point separating (i.e., if $k, h \in K$ are different, then $\{k, h\} \cap U$ is a singleton for at least one $U \in \mathcal{U}$) and each $\mathcal{U}_n$ is point finite (i.e., for every $k \in K$ the family $\{U \in \mathcal{U}_n : k \in U\}$ is finite);
  (vi) There exist a set Γ and sets $\Gamma_n \subset \Gamma$, $n \in \mathbb{N}$, with $\bigcup_{n=1}^\infty \Gamma_n = \Gamma$, and a continuous injective mapping $\Phi : K \to [0, 1]^\Gamma$ such that for every $k \in K$ and for every $n \in \mathbb{N}$ the set $\mathrm{supp} \, \Phi(k) \cap \Gamma_n$ is finite.

Proof. (i)$\Rightarrow$(ii). Let K be a compact set in (V, w), where V is a Banach space. By going to a subspace if necessary, we may, and do, assume that sp$K$ is dense in V. Then V is WCG. Now, unfortunately, we must go to Chapter 7 and use a deep Theorem 7.2.2(ii). Thus we get a set Γ and a linear continuous injective mapping $T: V \to c_0(\Gamma)$. Because T is weak-to-weak continuous, (ii) follows.

(ii)$\Rightarrow$(iii). Assume that K is a compact lying in $(c_0(\Gamma), w)$ with some Γ. Then K is norm bounded by the Banach–Steinhaus theorem. Thus, without loss of generality, we may, and do, assume that $\|k\| \leq \frac{1}{2}$ for all $k \in K$. For $\gamma \in \Gamma$ we define $\pi_\gamma : K \to \mathbb{R}$ as $\pi_\gamma(k) = k(\gamma)$, where $k = \{k(\alpha) : \alpha \in \Gamma\} \in K$. Clearly π_γ are continuous, that is, they belong to $C(K)$. It is also obvious that the family $\{\pi_\gamma : \gamma \in \Gamma\}$ separates the points of K, that is, for every $k, k' \in K$, with $k \neq k'$, there is $\gamma \in \Gamma$ such that $\pi_\gamma(k) \neq \pi_\gamma(k')$. Let L be the set consisting of the constant function equal to 1 on K and of all finite products of π_γ. (The γ's may repeat in such products.) Then, according to the Stone–Weierstrass theorem, sp$L$ is dense in $C(K)$. We shall show that L is weakly compact. For this, by the Eberlein–Šmulyan theorem, it is enough to show that every one-to-one sequence $\{g_n\}$ in L converges weakly to the zero function. Actually, by F. Riesz's representation theorem, Hahn's decomposition theorem [DS, Corollary III.4.11], and Lebesgue's dominated convergence theorem, it is enough to show that $g_n(k) \to 0$ for every $k \in K$. So fix $k \in K$ and $\epsilon > 0$ and let $n \in \mathbb{N}$ be such that $|g_n(k)| > \epsilon$. We recall that g_n is of the form $\prod_{\gamma \in F} \pi_\gamma^{m_\gamma}$, where F is a finite subset of Γ and $m_\gamma \in \mathbb{N}$. Since $\|\pi_\gamma\| \leq \frac{1}{2} < 1$, the set F must lie in $\{\alpha \in \Gamma : |\pi_\alpha(k)| > \epsilon\}$, which is equal to the set $\{\alpha \in \Gamma : |k(\alpha)| > \epsilon\}$. And this set is finite since $k \in c_0(\Gamma)$. It follows we have only finitely many different possibilities for the set F. Further, fix $\gamma \in F$. As $\|\pi_\gamma\| \leq \frac{1}{2}$, we have $(\frac{1}{2})^{m_\gamma} > \epsilon$. Thus m_γ are bounded by a fixed number, depending only on ϵ. By putting together all these facts and recalling that our sequence was one-to-one, we can conclude that $|g_n(k)| > \epsilon$ holds only for finitely many n's.

(iii)$\Rightarrow$(iv) is contained in Theorem 1.2.3.

(iv)$\Rightarrow$(i) is trivial since K can be understood as a subspace of $(B_{C(K)^*}, w^*)$.

(ii)$\Rightarrow$(v). Assume that $K \subset (c_0(\Gamma), w)$. Let $\{r_n\}$ be an enumeration of all nonzero rational numbers and put

$$U_\gamma^n = \{k \in K : r_n k(\gamma) > 1\}, \qquad n \in \mathbb{N}, \qquad \gamma \in \Gamma.$$

Clearly, each such U_γ^n is open and F_σ. Put further

$$\mathcal{U}_n = \{U_\gamma^n : \gamma \in \Gamma\}, \qquad n \in \mathbb{N},$$

and $\mathcal{U} = \bigcup_{n=1}^\infty \mathcal{U}_n$. Take any $k, h \in K$, $k \neq h$. We find $\gamma \in \Gamma$ such that $k(\gamma) \neq h(\gamma)$, say $k(\gamma) > h(\gamma)$. If $k(\gamma) > 0 \geq h(\gamma)$, then $k \in U_\gamma^n$, $h \notin U_\gamma^n$ for any $n \in \mathbb{N}$ satisfying $r_n > 1/k(\gamma)$. If $h(\gamma) > 0$, we find $n \in \mathbb{N}$ such that $r_n k(\gamma) > 1 > r_n h(\gamma)$; so $k \in U_\gamma^n$, $h \notin U_\gamma^n$. The case when $0 \geq k(\gamma)$ can be treated in a similar way. This shows that our family $\mathcal{U}$ is point separating. Fix now any $k \in K$ and any $n \in \mathbb{N}$. Then

$$\{U \in \mathcal{U}_n : \ k \in U\} = \{U^n_\gamma : \ k \in U^n_\gamma\}$$
$$= \{U^n_\gamma : \ r_n k(\gamma) > 1\} \subset \{U^n_\gamma : \ |k(\gamma)| > 1/|r_n|\}$$

and the last set is finite. Therefore $\mathcal{U}_n$ is point finite.

(v)$\Rightarrow$(vi). Let $\mathcal{U} = \bigcup_{n=1}^\infty \mathcal{U}_n$ be from (v). For every $U \in \mathcal{U}$ let $f_U : K \to [0, 1]$ be a continuous function such that $f_U(k) > 0$ if and only if $k \in U$; the existence of such a function is guaranteed by [En, Theorem 3.1.9, Corollary 1.5.12]. Define then $\Phi : K \to [0, 1]^\mathcal{U}$ by

$$\Phi(k)(U) = f_U(k), \qquad k \in K, \qquad U \in \mathcal{U}.$$

Clearly, Φ is continuous. It is injective as $\mathcal{U}$ separates the points of K. Take any $k \in K$ and any $n \in \mathrm{IN}$. Then

$$\mathrm{supp}\ \Phi(k) \cap \mathcal{U}_n = \{U \in \mathcal{U}_n : \ \Phi(k)(U) \neq 0\}$$
$$= \{U \in \mathcal{U}_n : \ f_U(k) > 0\} = \{U \in \mathcal{U}_n : \ k \in U\},$$

and the last set is known to be finite by (v). This proves (vi).

(vi)$\Rightarrow$(ii). Let Γ, Γ_n and Φ be as in (vi). Define $\Psi : [0, 1]^\Gamma \to [0, 1]^\Gamma$ by

$$\Psi(x)(\gamma) = \frac{1}{n} x(\gamma) \quad \text{if} \quad \gamma \in \Gamma_n \backslash \bigcup_{i=1}^{n-1} \Gamma_i, \quad n = 1, 2, \ldots, \quad x \in [0, 1]^\Gamma.$$

Clearly Ψ is continuous and one-to-one. Moreover $(\Psi \circ \Phi)(K) \subset c_0(\Gamma)$. $\qquad\square$

Theorem 1.2.5.

For a Banach space V the following assertions are equivalent:

  (i) V is WCG;

  (ii) There exist a set $\Gamma \neq \emptyset$ and a linear, continuous, weak*-to-weak continuous, and one-to-one mapping $S : V^* \to c_0(\Gamma)$;

  (iii) There exists a weakly compact set $K \subset V$ with the only weak accumulation point 0 and such that $\mathrm{sp}K$ is dense in V.

Proof. (i)$\Rightarrow$(ii). By Theorem 1.2.3, there is a reflexive space Y and a linear continuous mapping $T : Y \to V$ with $\overline{TY} = V$. Thus $T^* : V^* \to Y^*$ is weak*-to-weak continuous and injective. Now Y^* being reflexive, hence Vašák, there exists, by Theorem 7.2.2(ii), a linear continuous (hence weak-to-weak continuous) and injective mapping $R : Y^* \to c_0(\Gamma)$ for some set Γ. Now it is enough to put $S = R \circ T^*$.

(ii)$\Rightarrow$(iii). Let Γ and S be as in (ii) and let e_γ, $\gamma \in \Gamma$, be the canonical unit vectors in $\ell_1(\Gamma)$. They form a weak* relatively compact set with the only weak* accumulation point 0. Thus $K := \{S^* e_\gamma : \ \gamma \in \Gamma\} \cup \{0\}$ is a compact set in (V^{**}, w^*) and the continuity property of S guarantees that K lies in V; see [DS,

Theorem V.3.9]. (We identify V with its canonical image in V^{**}.) Hence K is a weakly compact set in V. We can easily verify that 0 is the only weak accumulation point of K. Finally, assume that $\operatorname{sp} K \neq V$. Then there exists $0 \neq v^* \in V^*$ with $v^*|_{\operatorname{sp} K} = 0$. Thus for all $\gamma \in \Gamma$ we have $0 = \langle S^* e_\gamma, v^* \rangle = \langle e_\gamma, S v^* \rangle$. So $S v^* = 0$, $v^* = 0$, a contradiction.

(iii)$\Rightarrow$(i) is trivial.   □

A proof that WCG spaces are weak Asplund is postponed to the next section.

1.3. ASPLUND GENERATED SPACES AND THEIR SUBSPACES

Definition 1.3.1. A Banach space V is called *Asplund generated* if there exists an Asplund space Y and a linear continuous mapping $T : Y \to V$ such that TY is dense in V.

Trivially, Asplund generated spaces are isomorphically invariant.

According to Theorem 1.1.2(ii), we can see that a Banach space V is Asplund generated if and only if there is an Asplund space Y such that Y is a dense subset of V and the corresponding inclusion mapping is continuous.

Trivially, Asplund spaces are Asplund generated. Also WCG spaces are Asplund generated according to Theorem 1.2.3.

In what follows, we shall focus on subspaces of Asplund generated spaces. This class is strictly larger than that of Asplund generated spaces; see Section 1.6.

To prove the next theorem, we shall need the following fact.

Proposition 1.3.2.

Let Z be a Baire space and let $\mathcal{U}$ be a family of open subsets in Z such that their union is dense in Z. Let Ω be a subset of Z such that for any $U \in \mathcal{U}$ the set $U \cap \Omega$ is residual in U. Then the whole Ω is residual in Z.

Proof. Let us well order the family $\mathcal{U}$, so that we can write $\mathcal{U} = \{U_\xi : \xi \in [0, \xi_0)\}$, where $[0, \xi_0)$ is an interval of ordinal numbers; this is guaranteed by Zermelo's theorem [En, p. 22]. Define the sets

$$W_0 = U_0, \qquad W_\xi = U_\xi \setminus \overline{\bigcup_{\eta < \xi} U_\eta}, \qquad \xi \in (0, \xi_0).$$

We shall show that the set $W := \bigcup_{\xi < \xi_0} W_\xi$ is dense in Z. So take any nonempty open set U in Z. Let ξ be the first ordinal from $[0, \xi_0)$ for which $U \cap U_\xi \neq \emptyset$. Then $U \cap \bigcup_{\eta < \xi} U_\eta = \emptyset$ and so $U \cap W_\xi \neq \emptyset$. Hence $U \cap W \neq \emptyset$ and the density of W is proved. Now fix any ξ in $[0, \xi_0)$. As $\Omega \cap U_\xi$ is, by assumption, residual in U_ξ, there are open sets $U_\xi^n \subset U_\xi$, $\overline{U_\xi^n} \supset U_\xi$, $n = 1, 2, \ldots$, such that $\Omega \cap U_\xi \supset \bigcap_{n=1}^\infty U_\xi^n$. Then $U_\xi^n \cap W_\xi$ are open and dense in W_ξ and also $\Omega \cap W_\xi \supset \bigcap_{n=1}^\infty (U_\xi^n \cap W_\xi)$. Performing this for each ξ, put $U^n = \bigcup \{U_\xi^n \cap W_\xi : \xi \in [0, \xi_0)\}$, $n = 1, 2, \ldots$. Clearly U^n are dense in Z. Moreover, observing that $W_\xi \cap W_\eta = \emptyset$ if $\xi \neq \eta$, we conclude that

$$\bigcap_{n=1}^\infty U^n = \bigcup \left\{ W_\xi \cap \bigcap_{n=1}^\infty U^n : \xi \in [0, \xi_0) \right\}$$

$$= \bigcup \left\{ \bigcap_{n=1}^\infty (U_\xi^n \cap W_\xi) : \xi \in [0, \xi_0) \right\} \subset \bigcup \{\Omega \cap W_\xi : \xi \in [0, \xi_0)\} \subset \Omega.$$

This shows that the set Ω is residual in Z. $\square$

Theorem 1.3.3.

Subspaces of Asplund generated spaces are weak Asplund.

Proof. Let Z be a subspace of some $V = \overline{TY}$, where Y is an Asplund space and $T : Y \to V$ is linear continuous. Let $f : Z \to \mathbb{R}$ be a continuous convex function. According to [Ph, Proposition 1.6], f is locally Lipschitz on Z. Let $\mathcal{U}$ denote the family of all open subsets U of Z such that the restriction $f_{|U}$ is Lipschitz; then $\bigcup \{U : U \in \mathcal{U}\} = Z$. Fix $U \in \mathcal{U}$ and let L be a Lipschitz constant of f on U. By Proposition 1.3.2, it will be enough to show that f is Gâteaux differentiable at the points of a set residual in U.

For $n, p \in \mathbb{N}$ put

$$A_n^p = \left(nTB_Y + \frac{1}{p} B_V \right) \cap Z$$

and let U_n^p be the set of all $z \in U$ such that there exists an open set $\Omega \ni z$ such that the A_n^p-diameter of $\partial f(\Omega) := \bigcup \{\partial f(z) : z \in U\}$ is less than $3L/p$; that is,

$$\sup \langle \partial f(\Omega) - \partial f(\Omega), A_n^p \rangle = \sup \left\{ \langle \xi - \eta, a \rangle : \xi, \eta \in \partial f(\Omega), \ a \in A_n^p \right\} < \frac{3L}{p}.$$

Fix $p, n \in \mathbb{N}$. Clearly, U_n^p is an open set. We shall show that U_n^p is dense in U. To do this, we shall need the following claim: *For every $M \subset LB_{Z^*}$ there are $0 \neq e \in Z$ and $\alpha > 0$ such that the weak* slice*

$$S(M, e, \alpha) = \{\xi \in M : \langle \xi, e \rangle > \sup \langle M, e \rangle - \alpha\}$$

has A_n^p-diameter less than $3L/p$.

Assume the claim was already proved and let $\emptyset \neq W \subset U$ be any open set. Then $\partial f(W) \subset LB_{Z^*}$ since f is Lipschitz on W with Lipschitz constant L. So, by the claim, there are $e \in Z$ and $\alpha > 0$ such that the A_n^p-diameter of $S(\partial f(W), e, \alpha)$ is less than $3L/p$. Find $z \in W$ and $\xi \in \partial f(z)$ such that

$\xi \in S(\partial f(W), e, \alpha)$. Let $t > 0$ be so small that $z + te \in W$. Then for any $\eta \in \partial f(z + te)$ we have, from the monotonicity of ∂f, that $\langle \eta, te \rangle \geq \langle \xi, te \rangle$. Thus

$$\inf \langle \partial f(z + te), e \rangle \geq \langle \xi, e \rangle > \sup \langle \partial f(W), e \rangle - \alpha.$$

Now, ∂f is norm-to-weak* upper semicontinuous; hence, there is an open set $z + te \in \Omega \subset W$ such that

$$\inf \langle \partial f(\Omega), e \rangle > \sup \langle \partial f(W), e \rangle - \alpha.$$

It follows $\partial f(\Omega) \subset S(\partial f(W), e, \alpha)$. Hence A_n^p-diam $\partial f(\Omega) < 3L/p$, and therefore, $\Omega \subset U_n^p \cap W$. This proves the density of U_n^p in U.

Thus the set $\bigcap \{U_n^p : n, p \in \mathbb{IN}\}$ is residual in U. Fix any z in this intersection. We shall show that $\partial f(z)$ is a singleton, which is equivalent to the Gâteaux differentiability of f at z. This way we shall get that Z is a weak Asplund space. Take any $e \in Z$ and any $\epsilon > 0$. We find $p \in \mathbb{IN}$ satisfying $3L/p < \epsilon$ and then we take $n \in \mathbb{IN}$ such that $A_n^p \ni e$; this is possible since $\bigcup_m A_m^p = Z$. Then

$$\sup \langle \partial f(z) - \partial f(z), e \rangle \leq A_n^p\text{-diam } \partial f(z) < \frac{3L}{p} < \epsilon.$$

It follows that $\langle \partial f(z), e \rangle$ is a singleton for each $e \in Z$ and hence $\partial f(z)$ is a singleton.

It remains to prove the claim. Take $p, n \in \mathbb{IN}$, $\emptyset \neq M \subset LB_{Z^*}$, and let $\widetilde{M}$ be its weak* closed convex hull. Let E be a weak* closed convex set in LB_{V^*}, minimal with respect to inclusion, and such that $\widetilde{M} = \{\xi_{|Z} : \xi \in E\}$; the existence of such an E can be easily shown by applying Zorn's lemma. Since Y is Asplund, by Theorem 1.1.1, there are $0 \neq y \in Y$ and $\beta > 0$ such that the weak* slice

$$S(T^*(E), y, \beta) = \{y^* \in T^*(E) : \langle y^*, y \rangle > \sup \langle T^*(E), y \rangle - \beta\}$$

has norm-diameter, that is B_Y-diameter, less than $L/(pn)$. Since $\overline{TY} = V$, we may and do assume that $Ty \neq 0$. Consider the weak* slice

$$S(E, Ty, \beta) = \{v^* \in E : \langle v^*, Ty \rangle > \sup \langle E, Ty \rangle - \beta\}.$$

Then

$$\begin{aligned}
(nTB_Y)\text{-diam } S(E, Ty, \beta) &= \sup \langle S(E, Ty, \beta) - S(E, Ty, \beta), nTB_Y \rangle \\
&= n \sup \langle S(T^*(E), y, \beta) - S(T^*(E), y, \beta), B_Y \rangle \\
&< n \frac{L}{pn} = \frac{L}{p}.
\end{aligned}$$

Now the set

$$M_1 = \left\{ \xi_{|Z} : \ \xi \in E \backslash S(E, Ty, \beta) \right\}$$

is weak* closed and convex, and by the minimality, $M_1 \neq \tilde{M}$. So, by the separation theorem, there are $0 \neq e \in Z$ and $\alpha > 0$ such that

$$S(\tilde{M}, e, \alpha) \cap M_1 = \emptyset.$$

Then $\xi \in S(E, Ty, \beta)$ whenever $\xi \in E$ and $\xi_{|Z} \in S(\tilde{M}, e, \alpha)$. Hence

$$A_n^p\text{-diam } S(M, e, \alpha) \leq A_n^p\text{-diam } S(\tilde{M}, e, \alpha)$$

$$= \sup \left\langle S(\tilde{M}, e, \alpha) - S(\tilde{M}, e, \alpha), \left(nTB_Y + \frac{1}{p}B_V \right) \cap Z \right\rangle$$

$$\leq \sup \left\langle S(E, Ty, \beta) - S(E, Ty, \beta), nTB_Y + \frac{1}{p}B_V \right\rangle$$

$$\leq \sup \left\langle S(E, Ty, \beta) - S(E, Ty, \beta), nTB_Y \right\rangle + \frac{2L}{p} < \frac{3L}{p}.$$

$\square$

Theorem 1.3.4.

Subspaces of WCG spaces are weak Asplund.

Proof. Put together Theorems 1.3.3 and 1.2.3. $\square$

Definition 1.3.5. The class of Banach spaces isomorphic to a subspace of an Asplund generated space is denoted by $\mathcal{A}$.

This class is stable with respect to several Banach space operations:

Theorem 1.3.6

(o) *If $Z \in \mathcal{A}$, then Z is isometric with a subspace of an Asplund generated space.*

(i) *If $V \in \mathcal{A}$ and Y is a subspace of V, then $Y \in \mathcal{A}$.*

(ii) *If $X \in \mathcal{A}$ and $T : X \to Z$ is a linear continuous mapping with $\overline{TX} = Z$, then $Z \in \mathcal{A}$; in particular, quotients of X are in $\mathcal{A}$.*

(iii) *If Γ is any nonempty set and $V_\gamma \in \mathcal{A}$ for each $\gamma \in \Gamma$, then $\left(\sum_{\gamma \in \Gamma} V_\gamma \right)_{c_0} \in \mathcal{A}$ and $\left(\sum_{\gamma \in \Gamma} V_\gamma \right)_{\ell_p} \in \mathcal{A}$ for each $1 < p < +\infty$.*

(iv) *If $V_n \in \mathcal{A}$ for each $n \in \mathbb{N}$, then $\left(\sum_{n=1}^\infty V_n \right)_{\ell_1} \in \mathcal{A}$.*

(v) *If $V_n \in \mathcal{A}$, $n = 1, 2, \ldots$, are subspaces of a Banach space V and $\mathrm{sp} \bigcup_{n=1}^\infty V_n$ is dense in V, then $V \in \mathcal{A}$.*

(vi) *If $Z \in \mathcal{A}$, and (Ω, Σ, μ) is a finite measure space, then $L_p(\Omega, \Sigma, \mu, Z) \in \mathcal{A}$ for each $1 \leq p < +\infty$.*

Proof. (o) Let i be an isomorphism of a Banach space $(Z, \|\cdot\|)$ into an Asplund generated space $(V, \|\cdot\|)$. We shall construct an equivalent norm $|\cdot|$ on V such that i will become an isometry. We may and do assume that

$$\|z\| \leq \|i(z)\| \leq c\|z\| \quad \text{for all} \quad z \in Z,$$

where $c > 0$ is a constant. Put

$$|v| = \inf\{\|v - i(z)\| + \|z\| : z \in Z\}, \qquad v \in V.$$

It is elementary to check that $|\cdot|$ is a seminorm on V. Further for $v \in V$ we have

$$\|v\| = \|v - i(0)\| + \|0\| \geq |v|$$
$$\geq \frac{1}{c} \inf\{\|v - i(z)\| + \|i(z)\| : z \in Z\} \geq \frac{1}{c}\|v\|$$

so that $|\cdot|$ is an equivalent norm on V. Moreover for every $z_0 \in Z$ we have

$$\|z_0\| = \inf\{\|z_0 - z\| + \|z\| : z \in Z\}$$
$$\leq \inf\{\|i(z_0) - i(z)\| + \|z\| : z \in Z\}$$
$$= |i(z_0)| \leq \|i(z_0) - i(z_0)\| + \|z_0\| = \|z_0\|,$$

that is, $\|z\| = |i(z)|$ for every $z \in Z$, which means that i is an isometry from $(Z, \|\cdot\|)$ into $(V, |\cdot|)$. And, trivially, the last space is Asplund generated. We also remark that (o) can be obtained from the proof of (ii) by putting $X = Z$. However, in this way, Z will be shown to be isometric with a subspace of another Asplund generated space.

(i) is clear from the definition of the class $\mathcal{A}$.

(ii) Take X in $\mathcal{A}$; then there exists an isomorphism, say i, of X into a subspace of an Asplund generated space V. We may and do assume that $\|x\| \leq \|i(x)\|$ for all $x \in X$. We find an Asplund space Y and a continuous linear mapping $k : Y \to V$ such that $\overline{k(Y)} = V$. Now let $T : X \to Z$ be a continuous linear mapping with $\overline{TX} = Z$. We have to show that $Z \in \mathcal{A}$. We define $j : Z \to \ell_\infty(B_{Z^*})$ by $j(z) = \{\langle z^*, z \rangle : z^* \in B_{Z^*}\}$, $z \in Z$; note that j is an isometry. We also define $S : V \to \ell_\infty(B_{Z^*})$ by

$$Sv = \{\langle \tilde{z}^*, v \rangle : z^* \in B_{Z^*}\}, \qquad v \in V,$$

where $\tilde{z}^* \in V^*$ is such that $\tilde{z}^* \circ i = T^* z^*$ and $\|\tilde{z}^*\| = \|z^*_{|iX}\|$; the existence of such a $\tilde{z}^*$ is guaranteed by the Hahn–Banach theorem. Then S is linear and continuous since $\|Sv\| \leq \sup\{\|\tilde{z}^*\| : z^* \in B_{Z^*}\}\|v\| \leq \|T\|\|v\|$. Moreover, we can check that the diagram

$$X \xrightarrow{\ i\ } V \xleftarrow{\ k\ } Y$$

$$\downarrow T \qquad \downarrow S$$

$$Z \xrightarrow{\ j\ } \ell_\infty(B_{Z^*})$$

commutes, that is, $S \circ i = j \circ T$. Now

$$j(Z) = j(\overline{TX}) = \overline{j \circ T(X)} = \overline{S \circ i(X)}$$
$$\subset \overline{SV} = \overline{S(\overline{k(Y)})} = \overline{(S \circ k)(Y)} \subset \ell_\infty(B_{Z^*}).$$

Here the last but one space is Asplund generated because Y is Asplund. Therefore $Z \in \mathcal{A}$. Since j is an isometry, by taking $X = Z$, we shall have also reproved (o).

(iii) Consider $(V_\gamma, \|\cdot\|_\gamma) \in \mathcal{A}$, $\gamma \in \Gamma$, and put $V = \left(\sum_{\gamma \in \Gamma}(V_\gamma, \|\cdot\|_\gamma)\right)_{c_0}$. We recall that V consists of $\{v_\gamma : \gamma \in \Gamma\} \in \prod_{\gamma \in \Gamma} V_\gamma$ such that $\{\|v_\gamma\|_\gamma : \gamma \in \Gamma\} \in c_0(\Gamma)$. For each $\gamma \in \Gamma$ we find a Banach space Z_γ, an isometry i_γ from V_γ into Z_γ (see (o)), an Asplund space Y_γ, and a linear continuous mapping $T_\gamma : Y_\gamma \to Z_\gamma$ such that $\overline{T_\gamma Y_\gamma} = Z_\gamma$. We may and do assume that $\|T_\gamma\| \leq 1$. Then, according to Theorem 1.1.2 (iv), $\left(\sum_{\gamma \in \Gamma} Y_\gamma\right)_{c_0}$ is Asplund and the mapping $\{y_\gamma\} \mapsto \{T_\gamma y_\gamma\}$ is linear and continuous and maps $\left(\sum_{\gamma \in \Gamma} Y_\gamma\right)_{c_0}$ onto a dense subset of $\left(\sum_{\gamma \in \Gamma} Z_\gamma\right)_{c_0}$. Hence, the last space is Asplund generated. Now, the assignment $\{v_\gamma\} \mapsto \{i_\gamma(v_\gamma)\}$ sends $\left(\sum_{\gamma \in \Gamma} V_\gamma\right)_{c_0}$ into $\left(\sum_{\gamma \in \Gamma} Z_\gamma\right)_{c_0}$ isometrically. Hence $\left(\sum_{\gamma \in \Gamma} V_\gamma\right)_{c_0} \in \mathcal{A}$. The proof that $\left(\sum_{\gamma \in \Gamma} V_\gamma\right)_{\ell_p}$ belongs to $\mathcal{A}$ for $1 < p < +\infty$ is quite analogous.

(iv) From (iii) we already know that $\left(\sum_{n=1}^\infty V_n\right)_{c_0} \in \mathcal{A}$. Moreover the mapping $\{v_n\} \mapsto \{2^{-n} v_n\}$ is linear and continuous and maps $\left(\sum_{n=1}^\infty V_n\right)_{c_0}$ onto a dense subset of $\left(\sum_{n=1}^\infty V_n\right)_{\ell_1}$ $\left(= \{\{v_n\} \in \prod_{n=1}^\infty V_n : \sum_{n=1}^\infty \|v_n\|_n < +\infty\}\right)$. Hence, by (ii), the last space is in $\mathcal{A}$.

(v) By (iv), we have that $\left(\sum_{n=1}^\infty V_n\right)_{\ell_1} \in \mathcal{A}$, and this space is sent onto a dense subset of V according to the recipe $\{v_n\} \mapsto \sum_{n=1}^\infty v_n$. Now (ii) ensures that $V \in \mathcal{A}$.

(vi) Let Z be isomorphic to a subspace of $V = Y$, where Y is an Asplund space, which is linearly and continuously embedded in V. First assume $1 < p < +\infty$. It is easy to check that $L_p(\Omega, \Sigma, \mu, Y)$ embeds linearly and continuously into a dense subset of $L_p(\Omega, \Sigma, \mu, V)$. Thus, $L_p(\Omega, \Sigma, \mu, Y)$ being Asplund by Theorem 1.1.2 (v), we can conclude that $L_p(\Omega, \Sigma, \mu, V)$ is Asplund generated. Therefore, $L_p(\Omega, \Sigma, \mu, Z)$, which is isomorphic to a subspace of $L_p(\Omega, \Sigma, \mu, V)$, must belong to $\mathcal{A}$. Now let $p = 1$. We already know that $L_2(\Omega, \Sigma, \mu, V)$ is Asplund generated. But the last space continuously embeds into a dense subset of $L_1(\Omega, \Sigma, \mu, V)$. Hence, by (ii), $L_1(\Omega, \Sigma, \mu, Z) \in \mathcal{A}$. $\square$

1.4. *ASPLUND SETS AND ASPLUND GENERATED SPACES*

Definition 1.4.1. A subset M of a Banach space V is called an *Asplund* set (in V) if it is nonempty and bounded and for every countable set $A \subset M$ the pseudometric space (V^*, ρ_A) is separable, where ρ_A is defined by

$$\rho_A(\xi, \eta) = \sup |\langle \xi - \eta, A \rangle|, \qquad \xi, \eta \in V^*.$$

Proposition 1.4.2.

Bounded sets of an Asplund space and weakly compact sets of any Banach space are Asplund sets.

Proof. The first statement immediately follows from Theorem 1.1.1. Let K be a weakly compact set in a Banach space V. Then, by Theorem 1.2.3, $K \subset T(B_Y)$ with Y reflexive and $T : Y \to V$ linear and continuous. Let $A \subset K$ be countable. We find a countable set $B \subset B_Y$ such that $A = T(B)$. Then (V^*, ρ_A) is isometric with (T^*V^*, ρ_B), which is a subspace of (Y^*, ρ_B). However, this last (pseudometric) space is separable since Y is reflexive. □

Lemma 1.4.3.

Let V be a Banach space.
(i) The family of Asplund sets in V is closed under taking subsets, closures, finite linear combinations, finite unions, closed absolutely convex hulls, and continuous linear images.
(ii) If $M_n \subset V$, $n = 1, 2, \ldots$, are Asplund sets and $\sum_{n=1}^{\infty} \sup\{\|v\| : v \in M_n\} < +\infty$, then $\sum_{n=1}^{\infty} M_n$ is an Asplund set.
(iii) If $M_n \subset V$, $n = 1, 2, \ldots$, are Asplund sets, then so is $\bigcap_{n=1}^{\infty} (M_n + (1/n)B_V)$.

Proof. The proof of (i) is left to the reader.
(ii) Let A be a countable subset of $\sum_{n=1}^{\infty} M_n$, where each M_n is an Asplund set in V. We find countable sets $A_n \subset M_n$ such that $A \subset \sum_{n=1}^{\infty} A_n$. For $n \in \mathbb{N}$ put $B_n = A_1 + \cdots + A_n$, $R_n = A_{n+1} + A_{n+2} + \cdots$. By (i), the set B_n is Asplund and countable. Hence we can find a countable set $C_n \subset B_{V^*}$ such that C_n is ρ_{B_n}-dense in B_{V^*}. Put $C = \bigcup_{n=1}^{\infty} C_n$; this is again a countable set. Take any $\xi \in B_{V^*}$. Since $A \subset B_n + R_n$, the ρ_A-distance of ξ from C is at most

$$2 \sup\{\|v\| : v \in R_n\}$$

for every $n \in \mathbb{N}$. By assumption, this number tends to 0 as $n \to +\infty$. Hence ξ lies in the ρ_A-closure of C. Now, it is clear that the set $\{mv^* : v^* \in C, \ m \in \mathbb{N}\}$ is countable and ρ_A-dense in V^*.

(iii) Take a countable set A in $\bigcap_{n=1}^{\infty}(M_n + (1/n)B_V)$, where each M_n is an Asplund set in V. We find countable sets $A_n \subset M_n$ such that $A \subset \bigcap_{n=1}^{\infty}(A_n + (1/n)B_V)$. Then there are countable sets $C_n \subset B_{V^*}$ such that each C_n is ρ_{A_n}-dense in B_{V^*}. Put $C = \bigcup_{n=1}^{\infty} C_n$. We observe that for $\xi, \eta \in B_{V^*}$ we have $\rho_A(\xi, \eta) \leq \rho_{A_n}(\xi, \eta) + 2/n$ for every $n \in \text{IN}$. Therefore C is ρ_A-dense in B_{V^*} and so the set $\{mv^* : v^* \in C, \ m \in \text{IN}\}$ is countable and ρ_A-dense in V^*. □

Theorem 1.4.4.

Let M be an Asplund set in a Banach space V. For $n = 1, 2, \ldots$ let $\|\cdot\|_n$ be Minkowski's functional of the set $2^n \text{co}(M \cup -M) + 2^{-n}B_V$, put $Z = \left(\sum_{n=1}^{\infty}(V, \|\cdot\|_n)\right)_{\ell_2}$, $Y = \{(v_1, v_2, \ldots) \in Z : v_1 = v_2 = \cdots\}$, and define $T : Y \to V$ by $T\big((v, v, \ldots)\big) = v$, $(v, v, \ldots) \in Y$. Then Y is an Asplund space and T is a continuous linear mapping with $T(B_Y) \supset M$ and with T^*V^* dense in Y^*.

In particular, a Banach space V is Asplund generated if and only if there is an Asplund set $M \subset V$ such that $\overline{\text{sp}}\, M = V$.

Proof. If $v \in M$, then $\|2^n v\|_n \leq 1$ for all $n = 1, 2, \ldots$, and so $\||(v, v, \ldots)\||^2 \leq \sum_{n=1}^{\infty} 2^{-2n} < 1$. (The norm $\||\cdot\||$ was defined prior to Lemma 1.2.2.) Also, if $v \in T(B_Y)$, then $v \in 2^n \text{co}(M \cup -M) + 2^{-n}B_V$ for each $n = 1, 2, \ldots$. Therefore

$$M \subset T(B_Y) \subset \bigcap_{n=1}^{\infty} \left(2^n \text{co}(M \cup -M) + 2^{-n}B_V\right),$$

and Lemma 1.4.3 guarantees that $T(B_Y)$ is an Asplund set in V. We shall show that Y is an Asplund space. Let X be a separable subspace of Y. We find, in B_X, a countable dense set A. Since $T(B_Y)$ is an Asplund set, $(V^*, \rho_{T(A)})$ is separable; that is, (T^*V^*, ρ_A) is separable. But Lemma 1.2.2 says that $\overline{T^*V^*} = Y^*$. Hence (Y^*, ρ_A) is separable. Therefore (Y^*, ρ_{B_X}) is separable and finally so is X^*. Then we can conclude, by Theorem 1.1.1, that Y is an Asplund space.

The proof of the last statement is then trivial. □

So far we did not present all features of the Asplund sets. This will be done in the next theorem. Let M be a nonempty bounded set in a Banach space V. For $\emptyset \neq B \subset V^*$ we put $\text{diam}_M B = \sup\{\langle v_1^* - v_2^*, m \rangle : v_1^*, v_2^* \in B, \ m \in M\}$. If $f : V \to \text{IR}$ is a convex continuous function, we say that f is M-*differentiable* at $v \in V$ if

$$\lim_{t \downarrow 0} \frac{1}{t}[f(v + th) + f(v - th) - 2f(v)] = 0$$

uniformly with respect to $h \in M$. Thus for $M = B_V$ we get the Fréchet differentiability, while the Gâteaux differentiability means that f is $\{h\}$-differentiable for every $h \in V$.

Theorem 1.4.5.

For a nonempty bounded set M in a Banach space V the following assertions are equivalent:

   (i) The set M is Asplund;

  (ii) There is an Asplund space Y and a linear continuous mapping $T : Y \to V$ such that $T(B_Y) \supset M$;

 (iii) Every continuous convex function on V is M-differentiable at each point of a dense G_δ subset of V;

 (iv) Every continuous convex function on V has at least one point of M-differentiability;

  (v) For every nonempty bounded set $B \subset V^*$ and for every $\epsilon > 0$ there are $0 \neq v \in V$ and $\alpha > 0$ such that

$$\operatorname{diam}_M S(B, v, \alpha) < \epsilon;$$

 (vi) For every separable subspace Z of V, with $Z \cap M \neq \emptyset$, for every nonempty bounded set $C \subset Z^*$ and for every $\epsilon > 0$ there are $0 \neq z \in Z$ and $\beta > 0$ such that

$$\operatorname{diam}_{Z \cap M} S(C, z, \beta) < \epsilon.$$

Proof. (i)$\Rightarrow$(ii) is contained in Theorem 1.4.4.

(ii)$\Rightarrow$(iii). Let $f : V \to \mathbb{R}$ be a convex continuous function. For $n = 1, 2, \ldots$ put

$$U_n = \left\{ v \in V : \ \operatorname{diam}_{T(B_Y)} \partial f(\Omega)) < \frac{1}{n} \quad \text{for some open set} \quad v \in \Omega \subset V \right\}.$$

Clearly, the sets U_n are open. We shall show that they are dense in V. So fix $n \in \mathbb{N}$ and a nonempty open set $W \subset V$. By diminishing it, if necessary, we may and do assume that $\partial f(W)$ is a bounded set [Ph, Proposition 1.6]. Since Y is Asplund, Theorem 1.1.1 gives $e \in Y$ and $\alpha > 0$ such that the weak* slice $S(T^*(\partial f(W)), e, \alpha)$ has norm-diameter less than $1/n$. Choose $y \in W$ and $\eta \in \partial f(y)$ such that $\langle T^*\eta, e \rangle > \sup \langle T^*(\partial f(W)), e \rangle - \alpha$. Take $t > 0$ so small that $y + te \in W$. Then, from the monotonicity of ∂f, we have

$$\inf \langle T^*(\partial f(y + te)), e \rangle \geq \langle T^*\eta, e \rangle > \sup \langle T^*(\partial f(W)), e \rangle - \alpha$$

and the norm-to-weak* upper semicontinuity of ∂f yields an open set $y + te \subset \Omega \subset W$ such that $\inf \langle T^*(\partial f(\Omega)), e \rangle > \sup \langle T^*(\partial f(W)), e \rangle - \alpha$. It then follows that $T^*(\partial f(\Omega)) \subset S(T^*(\partial f(W)), e, \alpha)$ and hence $\Omega \subset U_n$. Therefore $\overline{U_n} = V$ and $\bigcap_{n=1}^{\infty} U_n$ is a dense G_δ set.

Take any v in this intersection. It remains to show that f is M-differentiable at v. So fix $\epsilon > 0$. Take $n \in \mathbb{N}$ so that $1/n < \epsilon$. Since $v \in U_n$, there is an open set $v \in \Omega \subset V$ such that $\operatorname{diam}_{T(B_Y)} \partial f(\Omega) < 1/n$. We find $\delta > 0$ such that $u \in \Omega$ whenever $u \in V$ and $\|u - v\| < \delta$. Now take any $t \in (0, \delta/\|T\|)$ and any $y \in B_Y$. Then

$$\frac{1}{t} \left[f(v + tTy) + f(v - tTy) - 2f(v) \right] \leq \langle \xi_1 - \xi_2, Ty \rangle$$

$$\leq \operatorname{diam}_{T(B_Y)} \partial f(\Omega)) < \frac{1}{n} < \epsilon;$$

here we took ξ_1 in $\partial f(v + tTy)$ and ξ_2 in $\partial f(v - tTy)$. This means that f is $T(B_Y)$-differentiable and hence M-differentiable at v.

(iii)$\Rightarrow$(iv) is trivial.

Suppose that (iv) is satisfied and let B and ϵ be as in (v). We define $f : V \to \mathbb{R}$ by $f(v) = \sup\langle B, v \rangle$, $v \in V$. Clearly, f is a continuous convex function. Let $v \in V$ be a point of M-differentiability of f. We then find $t > 0$ so small that $f(v + tm) + f(v - tm) - 2f(v) < t\epsilon/2$ for all $m \in M$. Put $\alpha = t\epsilon/4$. Then for all $v_1^*, v_2^* \in S(B, v, \alpha)$ and for all $m \in M$ we have

$$t\langle v_1^* - v_2^*, m \rangle = \langle v_1^*, v + tm \rangle - \langle v_1^*, v \rangle + \langle v_2^*, v - tm \rangle - \langle v_2^*, v \rangle$$
$$< f(v + tm) + f(v - tm) - 2f(v) + 2\alpha < t\epsilon/2 + t\epsilon/2 = t\epsilon,$$

which means that $\operatorname{diam}_M S(B, v, \alpha) \leq \epsilon$ and so (v) is proved.

Assume (v) holds and let Z, C, and ϵ be as in (vi). Assume, for simplicity, that $C \subset B_{Z^*}$. We define $Q : B_{V^*} \to B_{Z^*}$ by $Qv^* = v^*|_Z$, $v^* \in B_{V^*}$. Let $B \subset B_{V^*}$ be a minimal (with respect to inclusion) weak* closed convex set such that $Q(B)$ is equal to the weak* closed convex hull $\overline{\operatorname{co}}^* C$ of C; such a B exists according to Zorn's lemma. By (v), there are $v \in V$ and $\alpha > 0$ such that $\operatorname{diam}_M S(B, v, \alpha) < \epsilon$. The minimality of B ensures that (the weak* closed convex set) $Q(B \backslash S(B, v, \alpha))$ is a proper subset of $\overline{\operatorname{co}}^* C$. Hence, by the separation theorem, there are $z \in Z$ and $\beta > 0$ such that $S(\overline{\operatorname{co}}^* C, z, \beta) \cap Q(B \backslash S(B, v, \alpha)) = \emptyset$. But this implies that

$$S(\overline{\operatorname{co}}^* C, z, \beta) \subset Q(S(B, v, \alpha)).$$

Now we can estimate

$$\operatorname{diam}_{Z \cap M} S(C, z, \beta) \leq \operatorname{diam}_{Z \cap M} S(\overline{\operatorname{co}}^* C, z, \beta) \leq \operatorname{diam}_{Z \cap M} Q(S(B, v, \alpha))$$
$$= \operatorname{diam}_{Z \cap M} S(B, v, \alpha) \leq \operatorname{diam}_M S(B, v, \alpha) < \epsilon.$$

Finally suppose (vi) holds. Let $A \subset M$ be a countable set. Put $Z = \overline{\operatorname{sp}} A$; it will be a separable subspace. We have to show that (V^*, ρ_A) is separable.

Assume this is not the case. Then there exist $\epsilon > 0$ and an uncountable set $L \subset B_{Z^*}$ such that $\rho_A(z_1^*, z_2^*) > \epsilon$ whenever $z_1^*, z_2^* \in L$. Let $\{z_n\}$ be a sequence contained in and dense in B_Z. Then the mapping $z^* \mapsto \{\langle z^*, z_n \rangle : n \in \mathrm{I\!N}\}$ sends (the compact space) (B_{Z^*}, w^*) continuously and injectively into $[-1, 1]^{\mathrm{I\!N}}$. Hence (B_{Z^*}, w^*) is a metrizable separable space and so the same holds for (L, w^*). Now, proceeding as in the proof of Theorem 1.3.3 (and using the same terminology), we get that there exists an uncountable set $\widetilde{L} \subset L$ with no weak* isolated points. However, by (vi), there are $0 \neq z \in Z$ and $\beta > 0$ such that $\mathrm{diam}_{Z \cap M} S(\widetilde{L}, z, \beta) < \epsilon$. Thus, in particular, $\rho_A(z_1^*, z_2^*) < \epsilon$ whenever $z_1^*, z_2^* \in S(\widetilde{L}, z, \beta)$. This means, according to a property of L, that $S(\widetilde{L}, z, \beta)$ is a singleton, which is impossible since $S(\widetilde{L}, z, \beta)$ is a relatively weak* open set in $\widetilde{L}$ and $\widetilde{L}$ has no weak* isolated points.   □

1.5. RADON--NIKODÝM COMPACTA AND ASPLUND GENERATED SPACES

Definition 1.5.1. A compact space K is called a *Radon–Nikodým* compact if there are an Asplund space Y and a continuous injection of K into (Y^*, w^*).

From Theorem 1.2.3, Lemma 1.4.3, and Theorem 1.1.3, we get:

Proposition 1.5.2.

Eberlein compacta and scattered compacta are Radon–Nikodým compacta.

Trivially, the unit interval $[0, 1]$ is an Eberlein (hence Radon–Nikodým) compact, which is not scattered. For a scattered non-Eberlein compact we can take the interval $[0, \omega_1]$ of ordinal numbers. The topology on this space is defined as follows: Every nonlimit $\alpha \in [0, \omega_1)$ is an isolated point and if $\lambda \in (0, \omega_1]$ is a limit ordinal, its neighborhood basis is formed by intervals $[\alpha, \lambda]$, $0 \leq \alpha \leq \lambda$. That $[0, \omega_1]$ is not an Eberlein compact can be shown as follows. Assume it is. By Theorem 1.2.4, we can think that $[0, \omega_1]$ is a subspace of $(c_0(\Gamma), w)$ for some set Γ; so each element of $[0, \omega_1]$ has a countable support in Γ. Then a not very complicated exhausting argument yields that ω_1 is a limit of a sequence $\{\alpha_n\} \subset [0, \omega_1)$. And this is impossible because $\sup_{n \in \mathrm{I\!N}} \alpha_n < \omega_1$. Finally, take $K = (B_{C([0,\omega_1])^*}, w^*)$. This is a Radon–Nikodým compact by definition and Theorem 1.1.3. Trivially, K is not scattered, and if K were Eberlein compact, then so would be $[0, \omega_1]$.

For a compact space K, let $M(K)$ denote the (Banach) space of regular Borel measures on K endowed with the norm

$$\|\mu\| = \sup\Big\{ \sum_{i=1}^{m} |\mu(A_i)| : \ A_i \subset K \text{ are mutually disjoint Borel sets}\Big\},$$

where $\mu \in M(K)$, we recall that, according to F. Riesz's representation theorem, the mapping which sends $\mu \in M(K)$ to the functional $f \mapsto \int f \,d\mu$, $f \in C(K)$, is a linear isometry between the space $(M(K), \|\cdot\|)$ and the space $C(K)^*$ endowed by the norm dual to the supremum norm on $C(K)$. In what follows we shall identify these two spaces. For $k \in K$ put $\delta_k(f) = f(k)$, $f \in C(K)$; thus $\delta_k \in C(K)^*$.

Lemma 1.5.3.

Let K be a compact space, let $B \subset B_{C(K)}$ be a nonempty set, and consider on $C(K)^*$ the pseudometric ρ_B defined as

$$\rho_B(\lambda, \mu) = \sup_{f \in B}\langle \lambda - \mu, f\rangle, \qquad \lambda, \mu \in C(K)^*.$$

Assume that the subspace $(\{\delta_k : k \in K\}, \rho_B)$ is separable. Then $(C(K)^*, \rho_B)$ is also separable.

Proof. We find a sequence $\{k_n\} \subset K$ such that the set $\{\delta_{k_n} : n \in \mathbb{N}\}$ is ρ_B-dense in $\{\delta_k : k \in K\}$. Denote $S = \operatorname{co}\{\pm\delta_{k_1}, \pm\delta_{k_2}, \dots\}$. This set is clearly ρ_B-separable. We shall show that S is ρ_B-dense in $B_{C(K)^*}$. Fix any $\epsilon > 0$ and take any $\mu \in B_{C(K)^*}$. Denote $D = \{\nu \in C(K)^* : \ \sup|\langle \nu, B\rangle| \leq \epsilon/2\}$. Clearly, D is a convex symmetric and weak* closed set. For $m = 1, 2, \dots$ we put

$$K_m = \Big\{ k \in K : \ \delta_k \in \big\{ \pm\delta_{k_1}, \dots, \pm\delta_{k_m}\big\} + D\Big\}.$$

It is easy to check that each K_m is a closed set. Moreover $K = \bigcup_{m=1}^{\infty} K_m$. Find $m \in \mathbb{N}$ so that $\|\mu\|(K\backslash K_m) < \epsilon/2$. Define $\nu(A) = \mu(A \cap K_m)$ for every Borel set $A \subset K$. Then $\nu \in B_{C(K)^*}$, and an easy calculation yields that $\|\mu - \nu\| \leq \|\mu\|(K\backslash K_m)$ $(< \epsilon/2)$ and so $\rho_B(\mu, \nu) < \epsilon/2$. We observe that ν lies in the weak* closed convex hull of $\{\pm\delta_k : k \in K_m\}$; denote it by E. Indeed, if not, then, by the separation theorem, there exists $f \in C(K)$, $\|f\| \leq 1$, such that $\langle \nu, f\rangle > \sup\langle E, f\rangle$. Here $\sup\langle E, f\rangle = \sup|f(K_m)|$ while $\langle \nu, f\rangle = \int f d\nu = \int_{K_m} f d\mu \leq \sup|f(K_m)|$, a contradiction. Thus, recalling the definition of K_m, we have

$$\nu \in E \subset \operatorname{co}\{\pm\delta_{k_1}, \dots, \pm\delta_{k_m}\} + D \subset S + D$$

and hence $\mu \in S + 2D$. Therefore the ρ_B-distance of μ from the set S is at most ϵ. Finally, $(C(K)^*, \rho_B)$ is separable since the ρ_B-separable set $\bigcup_{n=1}^{\infty} nS$ is ρ_B-dense in it. $\qquad\square$

Theorem 1.5.4.

For a compact space K the following assertions are equivalent:
  (i) K is a Radon–Nikodým compact;
  (ii) $C(K)$ is an Asplund generated space; and
  (iii) $\left(B_{C(K)^*}, w^*\right)$ is a Radon–Nikodým compact.

Proof. Assume (ii) is satisfied; so there exists an Asplund space Y and $T : Y \to C(K)$ with $\overline{TY} = C(K)$. Then $T^* : C(K)^* \to Y^*$ is injective and weak*-to-weak* continuous. Hence (iii) holds. (iii)⇒(i) is trivial because K canonically injects into $\left(C(K)^*, w^*\right)$.

Finally, let K be a Radon–Nikodým compact. For simplicity, we shall assume that $K \subset (Z^*, w^*)$, where Z is an Asplund space. As K is compact, an application of the Banach–Steinhaus theorem guarantees that K is norm bounded. By a multiplication, if necessary, we may and do assume that $K \subset B_{Z^*}$. Define $S : Z \to C(K)$ by $Sz(k) = \langle k, z \rangle$, $k \in K$, $z \in Z$. Clearly, S is linear and $S(B_Z) \subset B_{C(K)}$. Hence, by Lemma 1.4.3, $S(B_Z)$ is an Asplund set. Put $A = S(B_Z) \cup \{1\}$ and $M = \sum_{n=1}^{\infty} 2^{-n} A^n$. In the paragraph below, we shall show that A^2, A^3, ... are all Asplund sets. Then Lemma 1.4.3(ii) guarantees that M will still be an Asplund set. The set M also separates the points of K. Thus the Stone–Weierstrass theorem yields that $\overline{\mathrm{sp}}M = C(K)$ and Theorem 1.4.4 then says that $C(K)$ is an Asplund generated space.

It remains to show that A^n are Asplund sets. For this it will be enough to show that $A_1 \cdot A_2$ is an Asplund set whenever A_1 and A_2 are Asplund sets in $C(K)$. So let A_1, A_2 be such Asplund sets and denote $a = \sup\{\|f\| : f \in A_1 \cup A_2\}$. Let B be a countable subset of $A_1 \cdot A_2$. We have to show that $(C(K)^*, \rho_B)$ is separable. We find countable sets $C_1 \subset A_1$ and $C_2 \subset A_2$ such that $B \subset C_1 \cdot C_2$. Let us check that for $k, k' \in K$ we have

$$\rho_B(\delta_k, \delta_{k'}) = \sup_{f \in B} |f(k) - f(k')| \le \sup \left\{ |g(k)h(k) - g(k')h(k')| :\, g \in C_1,\ h \in C_2 \right\}$$

$$\le a \sup_{g \in C_1} |g(k) - g(k')| + a \sup_{h \in C_2} |h(k) - h(k')|$$

$$= a\rho_{C_1}(\delta_k, \delta_{k'}) + a\rho_{C_2}(\delta_k, \delta_{k'}) \le 2a\rho_{C_1 \cup C_2}(\delta_k, \delta_{k'}).$$

Now, since $C_1 \cup C_2$ is a countable subset of the Asplund set $A_1 \cup A_2$, the space $\left(C(K)^*, \rho_{C_1 \cup C_2}\right)$ is separable, and a fortiori, $\left(\{\delta_k : k \in K\}, \rho_{C_1 \cup C_2}\right)$ and $(\{\delta_k : k \in K\}, \rho_B)$ are separable. It then follows from Lemma 1.5.3 that $(C(K)^*, \rho_B)$ is also separable. Therefore, $A_1 \cdot A_2$ is an Asplund set. $\square$

Theorem 1.5.5.

A compact space K is the continuous image of a Radon–Nikodým compact if and only if $C(K) \in \mathcal{A}$, that is, $C(K)$ is isometric to a subspace of an Asplund generated space.

Proof. Assume that $K = \varphi(L)$, where φ is continuous and L is a Radon–Nikodým compact. Then the mapping $\widetilde{\varphi} : C(K) \to C(L)$ defined by $\widetilde{\varphi}(f)(l) = f(\varphi(l))$, $l \in L$, $f \in C(K)$, is a linear isometry into. Next, we recall that, by Theorem 1.5.4, $C(L)$ is an Asplund generated space.

Conversely, assume that $i : C(K) \to V$, where i is an isometry into, $V = \overline{TY}$, and Y is Asplund. Then (B_{V^*}, w^*) is a Radon–Nikodým compact and so is the set $(i^*)^{-1}(K) \cap B_{V^*}$. Now, an application of the Hahn–Banach theorem yields that $K = i^*\big((i^*)^{-1}(K) \cap B_{V^*}\big)$. Therefore K is the continuous image of a Radon–Nikodým compact. □

It is clear that if V is an Asplund generated space, then its dual unit ball with weak* topology is a Radon–Nikodým compact. A converse implication may be false; see Theorem 1.6.3. However, we have:

Theorem 1.5.6.

$V \in \mathcal{A}$ if and only if (B_{V^*}, w^*) is the continuous image of a Radon–Nikodým compact.

Proof. Let $V \in \mathcal{A}$. So, there is an Asplund generated space Y and an isometry i from V into Y; then $i^*(B_{Y^*}) = B_{V^*}$. Here i^* is weak*-to-weak* continuous and (B_{Y^*}, w^*) is a Radon–Nikodým compact. Conversely, if (B_{V^*}, w^*) is a continuous image of a Radon–Nikodým compact, Theorem 1.5.5 says that $C\big((B_{V^*}, w^*)\big) \in \mathcal{A}$. Hence $V \in \mathcal{A}$. □

To our knowledge, *there is no known example of a subspace V of an Asplund generated space such that (B_{V^*}, w^*) is not a Radon–Nikodým compact*. Or, more modestly: *It is unknown if a continuous image of a Radon–Nikodým compact is Radon–Nikodým.*

1.6. A SUBSPACE OF A WCG SPACE MAY NOT BE WCG OR ASPLUND GENERATED

The first example of this kind is due to Rosenthal [Ro]. He constructed a non-WCG subspace of $L_1(\Omega, \Sigma, \mu)$ for some large measure space (Ω, Σ, μ) with a finite measure μ. Note that $L_2(\Omega, \Sigma, \mu)$ injects continuously onto a dense subset of $L_1(\Omega, \Sigma, \mu)$, and hence this space is WCG.

The example that we present here is a subspace of $C(K)$, where K is an Eberlein compact. This example is due to Argyros [Arg] and is based on an "Eberleinization" of the Talagrand compact considered in Section 4.3.

Put $\Gamma = \mathbb{N}^{\mathbb{N}}$. For $n = 1, 2, \ldots$ let $\mathcal{A}_n$ be the family of all subsets $A \subset \Gamma$ with the following property: if $\sigma, \tau \in A$ and $\sigma \neq \tau$, then $\sigma|n = \tau|n$ and $\sigma(n+1) \neq \tau(n+1)$; here $\sigma|n$ means the sequence $\sigma(1), \ldots, \sigma(n)$. Thus every $A \in \mathcal{A}_n$ is at most countable. For $A \subset \Gamma$ let χ_A denote the characteristic function of A, that is, $\chi_A(\gamma) = 1$ if $\gamma \in A$ and $\chi_A(\gamma) = 0$ if $\gamma \in \Gamma \backslash A$. Then define

$$K = \bigcup_{n=0}^{\infty} \left\{ \frac{1}{n} \chi_A : A \in \mathcal{A}_n \right\}$$

and consider on K the topology inherited from the product topology on $[0, 1]^{\Gamma}$. It is an easy exercise to check that K is a compact space. If $x \in K$ is of form $x = (1/n)\chi_A$, with $A \in \mathcal{A}_n$, we define $Tx \in [0, 1]^{\Gamma}$ as

$$Tx(\sigma) = 2^{-n} 3^{-\sigma(n+1)} \chi_A(\sigma), \qquad \sigma \in \Gamma.$$

Then, using the definition of $\mathcal{A}_n$, we get that $Tx \in c_0(\Gamma)$. The verification of the continuity of T (with respect to the weak topology of $c_0(\Gamma)$) is straightforward. An easy argument reveals that T is also injective. So, we can conclude, by Theorem 1.2.4, that K is Eberlein compact.

We need two lemmas.

Lemma 1.6.1.

Let Γ_n, $n = 1, 2, \ldots$, be any sets in $\mathbb{N}^{\mathbb{N}}$ such that $\bigcup_{n=1}^{\infty} \Gamma_n = \mathbb{N}^{\mathbb{N}}$. Then there exist $n, m \in \mathbb{N}$ and an infinite set $A \in \mathcal{A}_m$ such that $A \subset \Gamma_n$.

Proof: We realize that $\mathbb{N}^{\mathbb{N}}$ with the product topology is completely metrizable and hence a Baire space. Thus there exists $n \in \mathbb{N}$ such that $\overline{\Gamma_n}$ has a nonempty interior. This means there is a finite sequence s, with length m, of positive integers such that $\{\sigma \in \mathbb{N}^{\mathbb{N}} : \sigma|m = s\} \subset \overline{\Gamma_n}$. Then for every $i \in \mathbb{N}$ there is $\sigma_i \in \mathbb{N}^{\mathbb{N}}$ such that $\sigma_i|m = s$, $\sigma_i(m+1) = i$, and $\sigma_i \in \Gamma_n$; put $A = \{\sigma_i : i \in \mathbb{N}\}$. Then clearly A is infinite, $A \in \mathcal{A}_n$, and $A \subset \Gamma_n$. $\qquad\square$

Lemma 1.6.2.

Let Γ, Δ be two infinite sets and $F : \Gamma \times \Delta \to \mathbb{R}$ a mapping with the following properties:

  (i) For every $\gamma \in \Gamma$ the set $\operatorname{supp} F(\gamma, \cdot) := \{\delta \in \Delta : F(\gamma, \delta) \neq 0\}$ is nonempty and at most countable.

  (ii) For every $\delta \in \Delta$ the set $\operatorname{supp} F(\cdot, \delta) := \{\gamma \in \Gamma : F(\gamma, \delta) \neq 0\}$ is at most countable.

Then there exist an ordinal λ and at most countable and pairwise disjoint sets $\emptyset \neq \Gamma_\alpha \subset \Gamma$ and $\emptyset \neq \Delta_\alpha \subset \Delta$, $\alpha < \lambda$, with $\bigcup_{\alpha<\lambda} \Gamma_\alpha = \Gamma$, such that for each pair (γ, δ) with $F(\gamma, \delta) \neq 0$ there exists $\alpha < \lambda$ such that $(\gamma, \delta) \in \Gamma_\alpha \times \Delta_\alpha$.

Proof. Let $A \subset \Gamma$, $B \subset \Delta$ be nonempty subsets such that if $F(\gamma, \delta) \neq 0$, then either $(\gamma, \delta) \in A \times B$ or $(\gamma, \delta) \in (\Gamma \backslash A) \times (\Delta \backslash B)$. We shall construct, by a standard "saturation" argument, at most countable sets $A_0 \subset A$, $B_0 \subset B$ such that $\bigcup_{\gamma \in A_0} \text{supp } F(\gamma, \cdot) \subset B_0$ and $\bigcup_{\delta \in B_0} \text{supp } F(\cdot, \delta) \subset A_0$: Take some $\gamma \in A$ and put $A_1 = \{\gamma\}$. Then put $B_1 = \text{supp } F(\gamma, \cdot)$, $A_2 = \bigcup\{\text{supp} F(\cdot, \delta) : \delta \in B_1\}$, $B_2 = \bigcup\{\text{supp} F(\gamma, \cdot) : \gamma \in A_2\}, \dots$. Finally, define $A_0 = \bigcup_{n=1}^\infty A_n$, $B_0 = \bigcup_{n=1}^\infty B_n$. These sets are at most countable, by assumption, and they have the desired properties.

Now we shall apply what we have done in the previous paragraph. First put $A = \Gamma$, $B = \Delta$ and let Γ_1, Δ_1 be the corresponding A_0, B_0, respectively. Assume now that for $\alpha < \beta$ we have already constructed mutually disjoint sets $\Gamma_\alpha \subset \Gamma$ and mutually disjoint sets $\Delta_\alpha \subset \Delta$ such that $\bigcup_{\gamma \in \Gamma_\alpha} \text{supp} F(\gamma, \cdot) \subset \Delta_\alpha$ and $\bigcup_{\delta \in \Delta_\alpha} \text{supp} F(\cdot, \delta) \subset \Gamma_\alpha$. If $\bigcup_{\alpha<\beta} \Gamma_\alpha = \Gamma$, we put $\lambda = \beta$ and stop the process. Otherwise we put Γ_β, Δ_β equal to the sets A_0, B_0 found in the first paragraph for $A = \Gamma \backslash \bigcup_{\alpha<\beta} \Gamma_\alpha$ and $B = \Delta \backslash \bigcup_{\alpha<\beta} \Delta_\alpha$ $\square$

Theorem 1.6.3.

There exists a non-WCG, even a non–Asplund generated, subspace of a WCG space of the type $C(K)$.

Proof. Let K be the compact space defined above. We already know that K is an Eberlein compact and hence $C(K)$ is WCG; see Theorem 1.2.4. For $\sigma \in \mathbb{N}^{\mathbb{N}}$ we define $\pi_\sigma : K \to \mathbb{R}$ by $\pi_\sigma(k) = k(\sigma)$, $k \in K$. Clearly $\pi_\sigma \in C(K)$. Let Y be the closed linear span of $\{\pi_\sigma : \sigma \in \mathbb{N}^{\mathbb{N}}\}$. We shall show that this Y is not WCG.

By contradiction, assume Y is WCG. Then, according to Theorem 1.2.5, there exists a weakly compact set $\{y_\delta : \delta \in \Delta\}$ in Y with the only weak accumulation point 0 and such that $\text{sp}\{y_\delta : \delta \in \Delta\}$ is dense in Y. We define $F : \mathbb{N}^{\mathbb{N}} \times \Delta \to \mathbb{R}$ as

$$F(\sigma, \delta) = y_\delta(\chi_{\{\sigma\}}), \qquad (\sigma, \delta) \in \mathbb{N}^{\mathbb{N}} \times \Delta.$$

(Note that $\chi_{\{\sigma\}}$ does belong to K.)

In what follows, we shall verify the assumptions of Lemma 1.6.2. So fix $\sigma \in \mathbb{N}^{\mathbb{N}}$. Assume that $F(\sigma, \delta) = 0$ for all $\delta \in \Delta$. Then $y(\chi_{\{\sigma\}}) = 0$ for all $y \in Y$ and hence, in particular, $\pi_\sigma(\chi_{\{\sigma\}}) = 0$, which is impossible. Hence $F(\sigma, \delta) \neq 0$ for some $\delta \in \Delta$. Take any $m \in \mathbb{N}$. Then the set $\{\delta \in \Delta : |F(\sigma, \delta)| > 1/m\}$ is finite since 0 is the only weak accumulation point of $\{y_\delta : \delta \in \Delta\}$. Then the set $\{\delta \in \Delta : F(\sigma, \delta) \neq 0\}$, being a countable union of such sets, must be at most countable. Now fix $\delta \in \Delta$. Again it is enough to check that

$\{\sigma \in \mathbb{N}^{\mathbb{N}} : |F(\sigma,\delta)| > 1/m\}$ is at most a finite set for each $m \in \mathbb{N}$; fix one $m \in \mathbb{N}$. Since $\{\pi_\sigma : \sigma \in \mathbb{N}^{\mathbb{N}}\}$ is a linearly dense set in Y, there is $y \in Y$ of form $y = \sum_{\sigma \in B} c_\sigma \pi_\sigma$, with a finite set $B \subset \mathbb{N}^{\mathbb{N}}$ and $c_\sigma \in \mathbb{R}$, $\sigma \in B$, such that $\|y_\delta - y\| < 1/m$. Now if $|F(\sigma,\delta)| > 1/m$, then surely $y(\chi_{\{\sigma\}}) \neq 0$ and so $\sigma \in B$.

By Lemma 1.6.2, there are mutually disjoint and at most countable sets $\Gamma_\alpha \subset \mathbb{N}^{\mathbb{N}}$, $\Delta_\alpha \subset \Delta$, $\alpha < \lambda$, with $\bigcup_{\alpha<\lambda} \Gamma_\alpha = \mathbb{N}^{\mathbb{N}}$, such that whenever $F(\sigma,\delta) \neq 0$, then $\sigma \in \Gamma_\alpha$ and $\delta \in \Delta_\alpha$ for some $\alpha < \lambda$. For each $\alpha < \lambda$ we enumerate the elements of Γ_α as $\{\sigma_1^\alpha, \sigma_2^\alpha, \ldots\}$. (This can be done also if Γ_α is finite.)

For $n \in \mathbb{N}$ put $\Gamma_n = \{\sigma_n^\alpha : \alpha < \lambda\}$; then, clearly, $\bigcup_{n=1}^\infty \Gamma_n = \mathbb{N}^{\mathbb{N}}$. Further, for $n,m \in \mathbb{N}$ we put $\Gamma_{nm} = \{\sigma \in \Gamma_n : |y_\delta(\chi_{\{\sigma\}})| > 1/m$ for some $\delta \in \Delta\}$. We checked above that $\bigcup_{m=1}^\infty \Gamma_{nm} = \Gamma_n$.

By Lemma 1.6.1, there are $n,m,l \in \mathbb{N}$ and an infinite (countable) set $A \in \mathcal{A}_l$ such that $A \subset \Gamma_{nm}$. Let us enumerate it as $\{\sigma_1, \sigma_2, \ldots\}$ and let $\alpha_i < \lambda$ be such that $\sigma_i \in \Gamma_{\alpha_i}$, $i \in \mathbb{N}$. Then $\sigma_i = \sigma_n^{\alpha_i}$ and there are $\delta_i \in \Delta$ (in fact, $\delta_i \in \Delta_{\alpha_i}$ by Lemma 1.6.2) with $|y_{\delta_i}(\chi_{\{\sigma_i\}})| > 1/m$. We also remark that if $i \neq j$, then $\sigma_i \neq \sigma_j$, $\alpha_i \neq \alpha_j$, and so $y_{\delta_i}(\chi_{\{\sigma_j\}}) = 0$.

Let B be a finite subset of A. Then for every $\sigma \in \mathbb{N}^{\mathbb{N}}$ we have (Note that $(1/l)\chi_B \in K$.)

$$\pi_\sigma\left(\frac{1}{l}\chi_B\right) = \frac{1}{l}\chi_B(\sigma) = \frac{1}{l}\sum_{\tau \in B}\chi_{\{\tau\}}(\sigma) = \frac{1}{l}\sum_{\tau \in B}\pi_\sigma(\chi_{\{\tau\}}).$$

Hence

$$y\left(\frac{1}{l}\chi_B\right) = \frac{1}{l}\sum_{\tau \in B}y(\chi_{\{\tau\}})$$

for all $y \in Y$. Now let $B_1 \subset B_2 \subset \cdots$ be a sequence of finite subsets of A with $\bigcup_{j=1}^\infty B_j = A$. Then $(1/l)\chi_{B_j} \to (1/l)\chi_A$, and so

$$\left|y_{\delta_i}\left(\frac{1}{l}\chi_A\right)\right| = \lim_{j\to\infty}\left|y_{\delta_i}\left(\frac{1}{l}\chi_{B_j}\right)\right| = \frac{1}{l}|y_{\delta_i}(\chi_{\{\sigma_i\}})| > \frac{1}{lm}$$

for all $i = 1,2,\ldots$. However, this is impossible because we know that $\{y_{\delta_i}\}$ is a relatively weakly compact set with the unique weak accumulation point 0. Therefore Y is WCG.

That Y is not an Asplund generated space will follow from Theorem 8.3.4(i). □

1.7. NOTES AND REMARKS

The concepts of Asplund and weak Asplund spaces were introduced by Asplund [As] under the names strong and weak differentiability spaces, respectively. The present names came from [NP] and [LP], respectively. It was shown in [As] that, *if V^* admits a dual LUR (strictly convex) norm, then V is Asplund (weak Asplund).* Thus, in particular, *if V^* (V) is separable, then V is Asplund (weak Asplund)* [DGZ2, Theorem II.2.6(i)]. More generally, *if V^* is WCG, then V is Asplund* [DGZ2, Theorem VII.2.7], and *if V is WCG, then it is weak Asplund* [DGZ2, Corollary VI.5.2., Theorem II.7.3]. For a different proof of the last fact see [B]. Another important sufficient condition for being an Asplund space, due to Ekeland and Lebourg, is the presence of an equivalent Fréchet smooth norm or at least a Fréchet smooth function with bounded nonempty support [EL]. Let us show the first statement. Assume that V has a Fréchet smooth norm $\| \cdot \|$. By Theorem 1.1.1, we may assume that V is separable. Let J denote the derivative of $\| \cdot \|$. Then J is a norm-to-norm continuous mapping from $V \backslash \{0\}$ to V^*, and the Bishop–Phelps theorem [Ph, Theorem 3.19] guarantees that $J(V)$ is norm-dense in the unit sphere of V^*. Hence V^* must be separable and so V is Asplund.

In Theorem 1.1.1, the efforts of many mathematicians are put together: Asplund [As], Gregory [Gi, p. 160], Namioka, Phelps [NP], Stegall [St1, St2].... It should be noted that (ii) can be reformulated as follows: Every separable subspace of V is Asplund. This means that *being an Asplund space is separably determined.* For a nice account on the Asplund spaces we refer to [Y].

There are plenty of other equivalent characterizations of the class of Asplund spaces. In order to present some of them we shall need some new definitions. Let Y be a Banach space. We say that Y is *dentable* if for every nonempty bounded set $A \subset Y$ and every $\epsilon > 0$ there exist $e^* \in Y^*$ and $\alpha > 0$ such that the slice $S(A, e^*, \alpha) := \{ y \in A : \langle e^*, y \rangle > \sup \langle e^*, A \rangle - \alpha \}$ has norm-diameter less than ϵ. Given $\epsilon > 0$ the *infinite ϵ-tree* is an infinite sequence $\{y_n\} \subset Y$ such that $y_n = \frac{1}{2}(y_{2n} + y_{2n+1})$ and $\|y_{2n} - y_n\| = \|y_{2n+1} - y_n\| \geq \epsilon$ for each $n \in$ IN. Finally, we say that the space Y has the the *Radon–Nikodým property* if for every finite measure space (Ω, Σ, μ) and for every μ-continuous vector measure $G : \Sigma \to Y$ of bounded variation there exists $g \in L_1(\Omega, \Sigma, \mu, Y)$ such that $G(A) = \int_A g \, d\mu$ for all $A \in \Sigma$. The equivalence of the Radon–Nikodým property and the dentability is the result of the combined effort of Rieffel, Maynard, Huff, Davis, and Phelps; see [Di1, Chapter 6] and [DU, Section V.3]. It is easy to check that a dentable space contains no bounded infinite ϵ-tree for any $\epsilon > 0$. The converse is in general false, as was shown by Bourgain and Rosenthal [BR]. In dual spaces we have: *For a Banach space V the assertions (i)–(iii) from Theorem 1.1.1 are equivalent to any of the following assertions:*

(iv) V^* is dentable;

(v) V^* contains no bounded infinite ϵ-tree for any $\epsilon > 0$; and

(vi) V^* has the Radon–Nikodým property.

Trivially (iii)$\Rightarrow$(iv). That (iv)$\Rightarrow$(v) and (iv)$\Leftrightarrow$(vi) follow from the general facts mentioned above. (ii)$\Rightarrow$(vi) goes back to Dunford and Pettis [Di1, Theorem

6.4.1; DU, Theorem III.3.1] and has been known for a long time. The implication (vi)⇒(ii) is a deep achievement due to Stegall [St1; DU, Section VII.2]. A simple proof of the implication (v)⇒(ii), which in fact substitutes Stegall's argument, is due to van Dulst and Namioka [DN]. However, it should be noted that Stegall's construction remains essential for other arguments, in particular for a theorem due to Huff and Morris that *a dual Banach space has the Radon–Nikodým property if and only if it has the Krein–Milman property* [DU, Section VII.2].

As regards to Theorem 1.1.2, (i) is from [DGZ2, Proposition VI.5.4]; (ii), (iii), and (iv) are from [NP]; and (v) is proved with help of [DU]. Here we needed the Radon–Nikodým property. The presented proofs, except (ii), use a separable reduction, thus differing from original ones. The assertion (iii) means that the Asplundness obeys the so-called three-space property. Theorem 1.1.3 is also from [NP].

There are a lot of other important results in the theory of Asplund spaces. An interested reader should consult the books [Ph, DGZ2, Bo, Gi, Di1, DU] and references therein.

A series of recent papers by Haydon provide a profound insight to the theory of Asplund spaces [Ha1–Ha6]. He considered locally compact scattered spaces of a special kind (namely trees), and Banach spaces $C_0(T)$ of continuous functions, vanishing "at infinity," on such spaces T, with the supremum norm. Note that each such $C_0(T)$ is an Asplund space by Theorem 1.1.3. Actually *such spaces admit a C^∞ smooth function with bounded nonempty support and even C^∞ partitions of unity. If T is a full uncountably branching tree with height ω_1, then $C_0(T)$ admits no equivalent Gâteaux smooth norm.* Using a dyadic tree with some special properties, Haydon also showed that *if a space has an equivalent Fréchet smooth norm, its quotient may not admit such a norm.* Another surprising discovery of Haydon is: *The following assertions about a tree T are equivalent: (i) $C_0(T)$ admits an equivalent Fréchet smooth norm; (ii) $C_0(T)$ admits an equivalent C^∞ smooth norm; and (iii) $C_0(T)$ admits an equivalent LUR norm.* One more result is: *There exists an Asplund space V, with $(B_{V^{**}}, w^*)$ Corson compact, which admits no renorming which is either strictly convex or Gâteaux smooth.* Everything mentioned above and more can be found in the survey article [Ha6]; see also [DGZ2, Sections VI.9 and VII.6]. The question of *whether every Asplund space admits a Fréchet smooth function with bounded nonempty support* still remains open.

For a historical account of the WCG spaces and for more results than those presented in Section 1.2, see [Di1, Chap. 5] and a basic paper of Amir and Lindenstrauss [AL]. Theorem 1.2.3 is known in the literature as an interpolation theorem. It is from [DFJP], where also a host of consequences can be found. What kind of interpolation can be recognized here? Consider two properties of a set in a Banach space: "to be weakly compact" and "to be a unit ball of a Banach space." After our interpolation we obtained a set sharing both of these properties, thus obtaining a reflexive space. In the proof of Theorem 1.4.4, due to Stegall [St3], we interpolated again; the only difference was that the first property was replaced by "to be an Asplund set." Let us mention the following factorization consequence of Theorem 1.2.3: *If* $T : Z \to V$ *is a linear weakly compact mapping, then it factors through a reflexive space* Y; *that is, there are bounded linear mappings* $T_1 : Z \to Y$ *and* $T_2 : Y \to V$ *such that* $T_2 \circ T_1 = T$ [DFJP; Di1, p. 162]. Theorems 1.2.4 (i)–(iv) and 1.2.5 are from [AL]. Theorem 1.2.4 (v), (vi) are due to Rosenthal [Ro].

The Asplund generated spaces, under the name GSG (Grothendieck–Šmulyan generated) spaces, were introduced and broadly studied by Ch. Stegall [St3]. He also showed that they are weak Asplund (thus reproving that WCG spaces are weak Asplund). For another proof of this fact see [FZ] or [Zi1, Zi2], where being weak Asplund is derived from a generic (sub)differentiability property of Lipschitz, in general nonconvex, functions.

Proposition 1.3.2 is called a localization principle [Ku, §10, V]. Theorem 1.3.3, in the more general setting of monotone operators, is due to Christensen and Kenderov [ChK]. They used a technique of minimal usco mappings. Our proof is rather a direct elaboration of the basic method of Kenderov [K2]; see the proof of [Ph, Theorem 2.30] and [FP]. A strengthening of [ChK] was recently obtained by Heisler: *A monotone operator on a subspace of an Asplund generated space is single-valued except at the points of a* σ-*cone supported set* [H]; he used ideas of Zajíček [Za]. For a nonconvex generalization of Theorem 1.3.3, see [FP] and its predecessors [Zi1, Zi2, FZ]. Theorem 1.3.4 was first proved, using a renorming, by Asplund [As]. Arguments in the proof of Theorem 1.3.5 are mostly due to Stegall. Note that the property of being (a subspace of) an Asplund generated or a WCG space is not a three-space property; see Theorem 2.3.1.

When comparing Theorem 1.3.5(i) with Theorem 1.1.2(i) a natural question arises: *If* $V \in \mathcal{A}$ *and* $T : Y \to V$ *is continuous and linear with* $T^* V^*$ *dense in* Y^*, *is* $Y \in \mathcal{A}$? That such an assertion does not hold is demonstrated by the following example shown to us by Stegall. Let K be Talagrand's compact contructed in Section 4.3; then there are Eberlein compacta $K_n \subset K$ ($K_n = \{\chi_A : A \in \mathcal{A}_n\}$, where $\mathcal{A}_n$ are from Section 1.6) such that $K = \cup_{n=1}^{\infty} K_n$. Put $Y = C(K)$ and $V = \left(\sum_{n=1}^{\infty} C(H_n) \right)_{c_0}$, where $H_n = K_1 \cup \cdots \cup K_n$, and let $T : Y \to V$ be defined as $Tf = \left\{ (1/n) f_{|H_n} \right\}$, $f \in Y$. Clearly, T is a bounded linear mapping. By Theorem 1.2.4, each $C(H_n)$ is WCG. It is then easy to show that V is WCG and hence, by Theorem 1.2.3, Asplund generated. Now, Theorem 4.3.1 guarantees that Y is a weakly $\mathcal{K}$-analytic, hence a Vašák space, see Definition 7.1.5. If Y were in the class $\mathcal{A}$, then, by Theorem 8.3.7, Y would be isomorphic to a subspace of a WCG space. However, this is not true, see Theorem 4.3.1. So it remains to show that $T^* V^*$ is dense in Y^*. We recall that, according to F. Riesz' theorem, we may identify the space of regular Borel measures on K with $C(K)^*$. So fix $\mu \in Y^*$ and let $\epsilon > 0$ be given. Then there is $n \in \mathrm{IN}$ such that $\|\mu\|(K \backslash H_n) < \epsilon$. We note that V^* is isometric with $\left(\sum_{n=1}^{\infty} C(H_n)^* \right)_{\ell_1}$; we shall identify these two spaces. Let $v^* \in V^*$ be a sequence whose members are all 0 except the n-th one, which is equal to $n\mu_{|K_n}$. Then for

$f \in B_Y$ we have

$$\langle \mu - T^* v^*, f \rangle = \int f \, \mathrm{d}\mu - \langle v^*, Tf \rangle = \int_{K \backslash H_n} f \, \mathrm{d}\mu \leq \|f\| \|\mu\| (K \backslash H_n) < \epsilon.$$

Hence $\|\mu - T^* v^*\| \leq \epsilon$, which means that $\overline{T^* V^*} = Y^*$.

It should be noted that Stegall, influenced by Grothendieck's memoir [G], originally used another name and another, completely different definition for the Asplund sets: He defines in [St3] that a subset M of a Banach space has *Grothendieck–Šmulyan* property if it is bounded and for every finite measure space (Ω, Σ, μ) and for every linear continuous mapping $T : X \to L_\infty(\Omega, \Sigma, \mu)$ the set $T(M)$ is equimeasurable, that is, for every $\epsilon > 0$ there is $A \in \Sigma$ such that $\mu(A) > \mu(\Omega) - \epsilon$ and that the set $\{(Tx)_{|A} : x \in M\}$ is relatively norm-compact in $L_\infty(\Omega, \Sigma, \mu)$. That *M is an Asplund set if and only if M has Grothendieck–Šmulyan property* was proved in [St3]. A more direct proof of this is due to Fitzpatrick and J.J. Uhl, Jr.; see [Bo, Theorem 5.5.4].

Lemma 1.4.3 and Theorem 1.4.4 are from [St3]. The assertions of Theorem 1.4.5 can be found in [Bo, Chapter 5]; see also [Fi]. By putting $M = B_V$, we can see that this theorem contains Theorem 1.1.1. Techniques used in the proof of Theorem 1.4.5 are, of course, those from the proof of Theorem 1.1.1. An inspection of the proof of [NP, Lemma 3] reveals that there is one more equivalent assertion here: (v') *For every bounded set* $\emptyset \neq B \subset V^*$ *and for every* $\epsilon > 0$ *there exists a weak* open set* $W \subset V^*$ *such that* $W \cap B \neq \emptyset$ *and* $\mathrm{diam}_M(W \cap B) < \epsilon$.

The ideas about Radon–Nikodým compacta go (at least implicitly) back to Dunford, Pettis, Phillips, and Grothendieck. An explicit study of this subject can be found in [St3, N1, N2]. Let us mention at least one important equivalence: *A compact space K is Radon–Nikodým if and only if there is a lower semicontinuous metric ρ on K such that for every nonempty set $M \subset K$ and for every $\epsilon > 0$ there is an open set $\Omega \subset K$ such that the set $M \cap \Omega$ is nonempty and has ρ-diameter less than ϵ.* Lemma 1.5.3 comes from [St1, St3]. Its proof presented here was shown to us by W. B. Moors. Theorem 1.5.4 is stated in [St3, p. 111].

So far we have neglected one mathematician, O. I. Reinov, who has been working in the "Radon–Nikodým stuff" over the last thirty years and whose papers and results are not well known enough in the mathematical community. He came to study Radon–Nikodým compacta via Radon–Nikodým operators [Re1, Re2, Re4]. The concept of a measurable set in itself, which is what we called an Asplund set, was introduced in [Re3]. Also Radon–Nikodým sets are defined there. A result similar to our Theorem 1.4.4 can be found in [Re5]. Properties of the Radon–Nikodým sets from the point of view of integral representations were considered in [Re7]. For further information see [Bo, Chapter 5, §9].

The interrelationship between Asplund spaces and WCG spaces will be studied in Chapter 8.

Chapter Two

Properties of Gâteaux Differentiability Spaces and Weak Asplund Spaces

A Banach space is called a *Gâteaux differentiability* space if every continuous convex function on it is Gâteaux differentiable at the points of a dense set. Unlike our knowledge of Asplund spaces, our knowledge of Gâteaux differentiability spaces and weak Asplund spaces is very poor. This chapter gathers together a modest amount of known (to us) properties of these spaces. The first section considers spaces of general form while the second section deals with spaces of the type $C(K)$. In the third section there is a counterexample showing that the weak Asplundness is not a three-space property.

2.1. GENERAL GÂTEAUX DIFFERENTIABILITY SPACES AND WEAK ASPLUND SPACES

We shall need the following:

Lemma 2.1.1. *Let K be a compact space such that each nonempty closed subset $Y \subset K$ has a G_δ point y with respect to Y, that is, it can be written as $\{y\} = Y \cap \bigcap_{n=1}^{\infty} G_n$, where G_n are open sets in K. Then K is sequentially compact.*

Proof. Let $\{k_n\}$ be a sequence in K. For $m = 1, 2, \ldots$ let Y_m be the closure of $\{k_n : n \geq m\}$. Then $Y = \bigcap_{m=1}^{\infty} Y_m$ is precisely the set of all cluster points of the sequence $\{k_n\}$. By hypothesis, there are $y \in Y$ and open sets $G_n \subset K$, $n \in \mathbb{N}$, such that $\{y\} = Y \cap \bigcap_{n=1}^{\infty} G_n$. Since K is regular, we may in addition assume that $\overline{G_{n+1}} \subset G_n$ for each n. As $Y_m \cap G_n \neq \emptyset$ for all $m, n \in \mathbb{N}$, there is a sequence $n_1 < n_2 < n_3 < \cdots$ such that $k_{n_i} \in G_i$ for each $i \in \mathbb{N}$. If now z is a cluster point of $\{k_{n_i}\}$, then z is a cluster point of $\{k_n\}$ and so $z \in Y$. Clearly, such a z belongs to $\bigcap_{i=1}^{\infty} \overline{G_i}$, and hence $z = y$. This proves that y is the only cluster point of the sequence $\{k_{n_i}\}$. Therefore this sequence converges to y. $\qquad\square$

Theorem 2.1.2.

Let V be a weak Asplund space or even only a Gâteaux differentiability space. Then every bounded sequence in its dual V^* has a weak* convergent subsequence; that is, bounded sets in V^* are relatively weak* sequentially compact.

Proof. Put $K = (B_{V^*}, w^*)$ and let $Y \neq \emptyset$ be a weak* closed set in K. In order to apply Lemma 2.1.1, we need to find a G_δ point of Y. We define $\varphi : V \to \mathbb{R}$ as $\varphi(v) = \sup\langle Y, v \rangle$, $v \in V$. Clearly, φ is a continuous convex function. Hence, by the assumption, it is Gâteaux differentiable at some $v_0 \in V$; denote by ξ its derivative at v_0. For $n = 1, 2, \ldots$ consider the sets

$$G_n = \left\{ \eta \in Y : \langle \eta, v_0 \rangle > \varphi(v_0) - \frac{1}{n} \right\};$$

they are open and $\xi \in G_n$ as $\langle \xi, v_0 \rangle = \varphi(v_0)$. Take any η in $\bigcap_{n=1}^{\infty} G_n$. Then for all $h \in V$ we have

$$\langle \eta - \xi, h \rangle = \lim_{t \downarrow 0} \frac{1}{t} \left[\langle \eta, v_0 + th \rangle - \langle \eta, v_0 \rangle \right] - \langle \xi, h \rangle$$

$$\leq \lim_{t \downarrow 0} \frac{1}{t} \left[\varphi(v_0 + th) - \varphi(v_0) \right] - \langle \xi, h \rangle = 0.$$

Hence $\eta = \xi$ and $\bigcap_{n=1}^{\infty} G_n = \{\xi\}$, which means that $\xi \in Y$ and that ξ is a G_δ point in Y. Now Lemma 2.1.1 applies. $\qquad\square$

Using the just proved theorem we get that ℓ_∞ *is not a Gâteaux differentiability space and hence is not a weak Asplund space.* Indeed, consider a sequence $\{\xi_n\} \subset \ell_\infty^*$ defined by $\langle \xi_n, f \rangle = f(n)$, $f \in \ell_\infty$, $n = 1, 2, \ldots$. Assume ℓ_∞ is a Gâteaux differentiability space. Then there exists a weak* converging subsequence $\{\xi_{n_i}\}$. Define $f \in \ell_\infty$ by $f(n_i) = (-1)^i$, $i = 1, 2, \ldots$, and $f(n) = 0$ otherwise. Then $\langle \xi_{n_i}, f \rangle = (-1)^i$ and so $\{\xi_{n_i}\}$ cannot converge in the weak* topology.

Theorem 2.1.3.

Let $T : V \to Y$ be a linear continuous mapping between Banach spaces V and Y such that TV is dense in Y. Then, if V is a Gâteaux differentiability space, so is Y.

Proof. Let $\varphi : Y \to \mathbb{R}$ be a convex continuous function. Then so is $\varphi \circ T$. Since V is a Gâteaux differentiability space, there is a dense set $D \subset V$ such that $\varphi \circ T$ is Gâteaux differentiable at each point of D. Now TD is dense in Y, and it remains to show that φ is Gâteaux differentiable at all of the points of TD. So fix $v \in D$ and $h \in Y$. If there is $k \in V$ such that $Tk = h$, we have

$$\lim_{t \downarrow 0} \frac{1}{t} \left[\varphi(Tv + th) + \varphi(Tv - th) - 2\varphi(Tv) \right]$$

$$- \lim_{t \downarrow 0} \frac{1}{t} \left[(\varphi \circ T)(v + tk) + (\varphi \circ T)(v - tk) - 2(\varphi \circ T)(v) \right]$$

$$= 0.$$

Then, assume $h \in Y \setminus TV$. Fix an arbitrary $\epsilon > 0$ and find $k \in V$ such that $\|h - Tk\| < \epsilon$. Since φ is Lipschitz in a neighborhood of Tv, with Lipschitz constant L, say [Ph, Proposition 1.6], we have

$$(0 \leq) \lim_{t \downarrow 0} \frac{1}{t} \left[\varphi(Tv + th) + \varphi(Tv - th) - 2\varphi(Tv) \right]$$

$$\leq 2L\epsilon + \lim_{t \downarrow 0} \frac{1}{t} \left[(\varphi \circ T)(v + tk) + (\varphi \circ T)(v - tk) - 2(\varphi \circ T)(v) \right]$$

$$= 2L\epsilon.$$

Hence again $\frac{1}{t} \left[\varphi(Tv + th) + \varphi(Tv - th) - 2\varphi(Tv) \right] \to 0$ as $t \downarrow 0$. Therefore φ is Gâteaux differentiable at Tv and Y is a Gâteaux differentiability space. $\quad\square$

It is not known if the above fact has an analogue in weak Asplund spaces. This is the case if the mapping $T : V \to Y$ is surjective. For a proof of this assertion we need the following topological result:

Proposition 2.1.4.

Let $g : X \to Y$ be a continuous surjective and open mapping acting between regular spaces X and Y. Assume there exists a dense completely metrizable subspace D of X. Then there exists a completely metrizable subspace Ω of D such that g maps Ω onto $g(\Omega)$ homeomorphically and moreover $g(\Omega)$ is dense G_δ in Y. Hence every regular superspace $Z \supset Y$, with Y dense in Z, contains a completely metrizable dense G_δ subspace.

Proof. Let d be a complete metric generating the topology of D. For $A \subset D$ let diam A denote the diameter of A with respect to d. By induction, we can easily construct families $\mathcal{U}_0, \mathcal{U}_1, \dots$ of nonempty open sets in X such that $\mathcal{U}_0 = \{X\}$, and for $n = 1, 2, \dots$ the following conditions are satisfied:

(i) for every $U \in \mathcal{U}_n$ there is $W \in \mathcal{U}_{n-1}$ such that $\overline{U} \subset W$;

(ii) for every $U \in \mathcal{U}_n$ diam$(\overline{U} \cap D) < 1/n$;

(iii) if U_1, $U_2 \in \mathcal{U}_n$ and $U_1 \neq U_2$, then $g(U_1) \cap g(U_2) = \emptyset$; and

(iv) $\mathcal{U}_n$ is a maximal family of open sets in X satisfying (i)–(iii).

We shall show that each $\bigcup\{g(U) : U \in \mathcal{U}_n\}$ *is a dense set in* Y. Trivially, this is true for $n = 0$. Assume $n \in \mathbb{N}$ is the first number for which this is not so. Then there is an open nonempty set $G \subset Y$ such that $G \cap g(U) = \emptyset$ for all $U \in \mathcal{U}_n$. But we know there is $U_{n-1} \in \mathcal{U}_{n-1}$ such that $G \cap g(U_{n-1}) \neq \emptyset$. From the continuity of g there is a nonempty open set $U \subset X$ such that $g(U) \subset G$, $\overline{U} \subset U_{n-1}$, and diam$(\overline{U} \cap D) < 1/n$. But then the family $\mathcal{U}_n \cup \{U\}$ satisfies (i)–(iii) and is strictly larger than $\mathcal{U}_n$, which is a contradiction with (iv).

Put $\Omega = D \cap \bigcap_{n=1}^{\infty} \bigcup\{U : U \in \mathcal{U}_n\}$; this is a G_δ set. We shall show that $g(\Omega)$ *is dense in* Y. So fix an open nonempty set $G \subset Y$ and, from regularity of Y, find an open nonempty set $G_0 \subset Y$ with $\overline{G_0} \subset G$. By the previous paragraph, there is $U_1 \in \mathcal{U}_1$ such that $G_0 \cap g(U_1)$ is a nonempty set, and since g is an open mapping, this set is open. Hence, using the previous paragraph again, we can find $U_2 \in \mathcal{U}_2$ such that $G_0 \cap g(U_1) \cap g(U_2) \neq \emptyset$. Continuing in this way, we obtain for each $n = 1, 2, \ldots$ a set $U_n \in \mathcal{U}_n$ such that $G_0 \cap \bigcap_{n=1}^{m} g(U_n) \neq \emptyset$, $m = 1, 2, \ldots$. Fix for a moment one $n \in \mathbb{N}$ and find, by (i), a set $U \in \mathcal{U}_n$ such that $\overline{U_{n+1}} \subset U$. Then $g(U_n) \cap g(U) \supset g(U_n) \cap g(U_{n+1}) \neq \emptyset$ and, by (iii), we get $U = U_n$. Thus we have $\overline{U_{n+1}} \subset U_n$. Moreover, $G_0 \cap g(U_n) \neq \emptyset$, so that $g^{-1}(G_0) \cap U_n \neq \emptyset$. It follows that the sets $g^{-1}(\overline{G_0}) \cap \overline{U_n} \cap D$ are nonempty, closed in D, have d-diameter less than $1/n$, and form a nonincreasing sequence. Hence, by the Cantor theorem [En, Theorem 4.3.8], their intersection consists of one point, $x \in D$, say. We note that $x \in \bigcap_{n=1}^{\infty} \overline{U_n} = \bigcap_{n=1}^{\infty} U_n$. So $x \in \Omega$; $g(x) \in \overline{G_0} \cap g(\Omega) \subset G \cap g(\Omega)$ and the density of $g(\Omega)$ in Y is proved.

Now we remark that g *is injective on* Ω. In fact, let x_1, x_2 be in Ω and $x_1 \neq x_2$. We find $n \in \mathbb{N}$ so large that $d(x_1, x_2) > 1/n$ and further find $U_1, U_2 \in \mathcal{U}_n$ such that $U_1 \ni x_1$, $U_2 \ni x_2$. Then surely $U_1 \neq U_2$ and hence, by (iii), $g(U_1) \cap g(U_2) = \emptyset$. Thus $g(x_1) \neq g(x_2)$.

Further we shall show that *the restriction of g to Ω is open, that is, it sends relatively open sets in Ω to relatively open sets in $g(\Omega)$.* First observe that $g(\Omega \cap U) = g(\Omega) \cap g(U)$ if $U \in \mathcal{U}_n$. Indeed, take $U \in \mathcal{U}_n$ and $y \in g(\Omega) \cap g(U)$. Writing $y = g(x)$, with $x \in \Omega$, we have that $x \in W$ for some $W \in \mathcal{U}_n$. Hence $g(U) \cap g(W) \ni y$ and so $U = W$ by (iii). Therefore $x \in \Omega \cap U$ and $y = g(x) \in g(\Omega \cap U)$. Now let $W \subset X$ be any nonempty open set. We have to show that the set $g(\Omega \cap W)$ is relatively open in $g(\Omega)$. So take $y \in g(\Omega \cap W)$. Write $y = g(x)$, where $x \in \Omega \cap W$. Find $n \in \mathbb{N}$ so large that $d(x, z) > 2/n$ for all $z \in D \backslash (\Omega \cap W)$. Also, choose $U \in \mathcal{U}_n$ such that $x \in U$. Then $\Omega \cap U \subset \Omega \cap W$ and so $g(\Omega \cap W) \supset g(\Omega \cap U) = g(\Omega) \cap g(U)$. We observe that the last set is open in $g(\Omega)$ and contains y. Therefore $g(\Omega \cap W)$ is open in $g(\Omega)$.

Putting together the facts proved in the previous three paragraphs, we can conclude that g maps Ω homeomorphically onto $g(\Omega)$ and that this set is dense in Y. Note that Ω is a G_δ set in D. Thus Ω is a completely metrizable space [En, Theorem 4.3.23] and hence so is $g(\Omega)$.

The last statement is a standard fact. We include its proof for completeness. Denote, for simplicity, $C = g(\Omega)$. Let δ be a complete metric on C generating the topology inherited from Y. Consider a regular superspace $Z \supset Y$ such that Y is dense in Z. It remains to prove that C is a G_δ set in Z. For $n = 1, 2, \ldots$ put

$$G_n = \bigcup \left\{ W : \ W \subset Z, \ W \text{ is open and } \delta\text{-diam}(W \cap C) < \frac{1}{n} \right\}.$$

Clearly, $C \subset G_n$. So $\overline{G_n} = Z$ because we know that C is dense in Z. We shall show that $\bigcap_{n=1}^{\infty} G_n \subset C$. Take $z \in \bigcap_{n=1}^{\infty} G_n$. For each $n = 1, 2, \ldots$ find an open set $W_n \subset Z$ such that $W_n \ni z$, $\delta\text{-diam}(W_n \cap C) < 1/n$, and $\overline{W_{n+1}} \subset W_n$. From completeness of δ we know, by the Cantor theorem, that $\bigcap_{n=1}^{\infty} (\overline{W_n} \cap C)$ is a singleton, $\{y\}$, say [En, Theorem 4.3.8]. We claim $y = z$. Assume this is not true. Find then an open set $y \in W \subset Z$ with $z \notin \overline{W}$. As $\delta\text{-diam}(W_n \cap C) < 1/n$, there is $n \in \mathbb{N}$ so large that $W_n \cap C \subset W$. But C is dense in Z. So $W_n \subset \overline{W_n \cap C} \subset \overline{W}$ and hence $z \in \overline{W}$, a contradiction. We have thus proved that $\bigcap_{n=1}^{\infty} G_n = C$ and therefore C is a G_δ set in Z.   □

Theorem 2.1.5.

A quotient of a weak Asplund space is weak Asplund.

Proof. Assume T is a linear continuous mapping from a weak Asplund space V onto a Banach space Y. We have to show that Y is weak Asplund. We know that T is open. Let $f : Y \to \mathbb{R}$ be any convex continuous function. Since V is weak Asplund, there is a dense G_δ set $D \subset V$ such that $f \circ T$ is Gâteaux differentiable at every point of D. Now, since D is a G_δ subset of a Banach space, it follows that D is completely metrizable. Hence, according to Proposition 2.1.4, there is a set $\Omega \subset D$ such that $T(\Omega)$ is dense G_δ in Y. Now, as in the proof of Theorem 2.1.3, in fact more easily, we can show that f is Gâteaux differentiable at the points of $T(\Omega)$. Therefore Y is weak Asplund.   □

2.2. GÂTEAUX DIFFERENTIABILITY SPACES AND WEAK ASPLUND SPACES OF TYPE C(K)

Let K be a compact space and let $C(K)$ denote the Banach space of continuous functions on K endowed with the supremum norm $\| \cdot \|$. We define a function $\varphi_K : C(K) \to \mathbb{R}$ and a multivalued mapping $F_K : C(K) \to 2^K$ by

$$\varphi_K(x) = \sup x(K), \qquad x \in C(K),$$

$$F_K x = \{k \in K : x(k) = \varphi_K(x)\}, \qquad x \in C(K).$$

Clearly, φ_K is continuous and convex and $F_K x = \partial\varphi(x) \cap K$, where $\partial\varphi(x)$ is the usual subdifferential of φ at x. Further put

$$M_K = \{x \in C(K) : F_K x \text{ is a singleton}\}$$

and define a (single-valued) mapping $f_K : M_K \to F_K(M_K)$ by

$$f_K(x) = F_K x, \qquad x \in C(K).$$

We recall that a point in a topological space is called a G_δ *point* if it can be expressed as an intersection of countably many open sets.

Lemma 2.2.1.

For a compact space K we have:
  (i) The mapping F_K is a usco, that is, $\{x \in C(K) : F_K x \subset G\}$ is an open set whenever $G \subset K$ is open and $F_K x$ is a nonempty compact set for each $x \in C(K)$.
  (ii) For $x \in C(K)$ the set $F_K x$ is a singleton if and only if φ_K is Gâteaux differentiable at x.
  (iii) The set of G_δ points of K coincides with $F_K(M_K)$.
  (iv) The mappings F_K and f_K are open, that is, they map open sets onto open sets.

Proof. For simplicity, we shall write F, f, M instead of F_K, f_K, M_K, respectively.

(i) Clearly, Fx is a nonempty compact for each $x \in C(K)$. Let $Fx \subset G$, where $x \in C(K)$ and $G \subset K$ is an open set. Assume, by contradiction, that there are $x_n \in C(K)$, with $\|x_n - x\| \to 0$, and $k_n \in Fx_n$ such that $k_n \notin G$. Then

$$\sup x(K) \leq \sup x_n(K) + \|x_n - x\| = x_n(k_n) + \|x_n - x\| \leq x(k_n) + 2\|x_n - x\|.$$

Hence, if k is a cluster point of $\{k_n\}$, we get $\sup x(K) \leq x(k)$, so $k \in Fx \subset G$. But $k \notin G \supset Fx$ as $k_n \notin G$, a contradiction.

(ii) Fix some $x \in C(K)$ such that Fx is a singleton, $\{k_0\}$, say. Consider $h \in C(K)$ and let $\epsilon > 0$ be given. We find an open set $k_0 \in U \subset K$ such that $|h(k) - h(k_0)| < \epsilon/2$ whenever $k \in U$. Also, we choose $\delta > 0$ so small that $F(x + th) \subset U$ for $t \in (-\delta, \delta)$. Fix any $t \in (0, \delta)$ and find $k_t \in F(x + th)$, $k_t' \in F(x - th)$. Then

$$\frac{1}{t}\big[\varphi(x+th)+\varphi(x-th)-2\varphi(x)\big]$$

$$\leq \frac{1}{t}\big[(x+th)(k_t)-x(k_t)+(x-th)(k_t')-x(k_t')\big]$$

$$= h(k_t)-h(k_t') < \epsilon.$$

Thus φ is Gâteaux differentiable at x. Conversely, assume φ is Gâteaux differentiable at some $x \in C(K)$ and consider $k, k' \in Fx$. Then for every $y \in C(K)$ we have

$$y(k)-y(k') = \frac{1}{t}\big[(x+ty)(k)+(x-ty)(k')-x(k)-x(k')\big]$$

$$\leq \frac{1}{t}\big[\varphi(x+ty)+\varphi(x-ty)-2\varphi(x)\big] \to 0$$

as $t \downarrow 0$. Hence $k = k'$ by [En, Theorem 1.5.10].

(iii) Let $k_0 \in F(M)$. Then $\{k_0\} = Fx$ for some $x \in M$ and hence $\{k_0\} = \bigcap_{n=1}^{\infty}\{k \in K : x(k) > \varphi(x) - 1/n\}$, that is, k_0 is a G_δ point in K. Conversely, let k_0 be a G_δ point in K. Then $\{k_0\} = \bigcap_{n=1}^{\infty} H_n$, where $H_n \subset K$ are open sets. From compactness of K we find $x_n \in C(K)$ such that $0 \leq x_n(k) \leq 1 = x_n(k_0)$ for all $k \in K$ and $x_n(k) = 0$ for all $k \in K \backslash H_n$ [En, Theorem 1.5.10]. Put $x(k) = \sum_{n=1}^{\infty} 2^{-n} x_n(k)$, $k \in K$. Then $x(k_0) = 1$ and $0 \leq x(k) < x(k_0)$ whenever $k \in K \backslash \{k_0\}$. Clearly, x is continuous. It follows that $Fx = \{k_0\}$ and $x \in M$.

(iv) Fix $x \in C(K)$ and $\epsilon > 0$. Let $B(x, r)$ denote the open ball around x with diameter r. We claim that *there exists an open set $G_{x,\epsilon} \subset K$ such that*

$$Fx \subset G_{x,\epsilon} \subset F\big(B(x, 2\epsilon)\big),$$

and if $x \in M$, then

$$Fx \subset G_{x,\epsilon} \cap F(M) \subset F\big(B(x, 3\epsilon) \cap M\big).$$

Proof. Put

$$G_{x,\epsilon} = \{k \in K : x(k) > \sup x(K) - \epsilon\};$$

this set is open and contains Fx. Define $y \in C(K)$ by

$$y(k) = \min\{x(k) + \epsilon, \sup x(K)\}, \qquad k \in K;$$

then $\|y - x\| \leq \epsilon < 2\epsilon$. Thus $Fx \subset G_{x,\epsilon} \subset Fy \subset F\big(B(x, 2\epsilon)\big)$. Further assume $x \in M$ and choose any $k_0 \in G_{x,\epsilon} \cap F(M)$. So $k_0 = Fh$ for some $h \in M$; we may and do assume that $h \geq 0$ and $h(k_0) = 1$. Then

$$(y + \epsilon h)(k) < (y + \epsilon h)(k_0)$$

whenever $k \in K$ and $k \neq k_0$. Hence $y + \epsilon h \in M$. Moreover

$$\|y + \epsilon h - x\| \leq \|y - x\| + \epsilon\|h\| < 2\epsilon + \epsilon = 3\epsilon.$$

It follows that $G_{x,\epsilon} \cap F(M) \subset F\big(B(x, 3\epsilon) \cap M\big)$.

We recall that the (single-valued) mapping $f : M \to F(M)$ is defined by $f(x) = Fx$, $x \in M$. Now we are ready to prove (iv). Let $U \subset C(K)$ be an open nonempty set. For every $x \in U$ we find $\epsilon_x > 0$ such that $B(x, 3\epsilon_x) \subset U$. Then, by the claim, we have

$$F(U) = \bigcup_{x \in U} Fx \subset \bigcup_{x \in U} G_{x,\epsilon_x} \subset \bigcup_{x \in U} F(B(x, 2\epsilon_x)) \subset F(U),$$

and

$$f(U \cap M) = \bigcup \{Fx : x \in U \cap M\} \subset \bigcup \{G_{x,\epsilon_x} \cap F(M) : x \in U \cap M\}$$
$$\subset \bigcup \{F(B(x, 3\epsilon_x) \cap M) : x \in U \cap M\} \subset F(U \cap M) = f(U \cap M).$$

This means $F(U)$ is open in K and $f(U \cap M)$ is open in $F(M)$. $\qquad\square$

Proposition 2.2.2.

For a compact space K the following assertions are equivalent:

  (i) The set $\{x \in C(K) : F_K x$ is a singleton$\}$ is dense (contains a dense G_δ set);

  (ii) The set $\{x \in C(K) : \varphi_K$ is Gâteaux differentiable at $x\}$ is dense (contains a dense G_δ set);

  (iii) The set of G_δ points of K is dense (contains a dense G_δ completely metrizable set).

Proof. Assertion (i) is equivalent to (ii) according to Lemma 2.2.1(ii). In what follows we shall again drop the index K.

(i)⇒(iii). Assume first that M is dense in $C(K)$. By Lemma 2.2.1(iii), we have to show that $F(M)$ is dense in K. So fix an open nonempty set $G \subset K$. The compactness of K yields $x \in C(K)$ such that $0 \leq x(k) \leq 1$ for all $k \in K$, $x(k) = 0$ for $k \in K \backslash G$, and that $x(k_0) = 1$ for some $k_0 \in U$ [En, Theorem 1.5.10]. Clearly, $Fx \subset G$, and by Lemma 2.2.1(i), there is an open set $x \in U \subset C(K)$ such that $F(U) \subset G$. Now we recall that M is dense in $C(K)$. So $U \cap M$ is nonempty; take some y here. Then $Fy \subset G$ and therefore $F(M) \cap G \neq \emptyset$.

Further assume that M contains a dense G_δ set D. In Proposition 2.1.4 put $X = M$, $Y = f(M)$, $g = f$, and $Z = K$. Since $C(K)$ is a Banach space, D is completely metrizable [En, Theorem 4.3.23]. We also know that $f(M)$ is dense in K. So Proposition 2.1.4 applies. Thus K contains a dense G_δ completely metrizable subset. And, clearly, each point of such a set is a G_δ point of K.

(iii)$\Rightarrow$(i). Assume first $F(M)$ is dense in K and let $\emptyset \neq U \subset C(K)$ be open. By Lemma 2.2.1(iv), $F(U)$ is open. Hence $F(M) \cap F(U) \neq \emptyset$; choose $h \in M$ and $x \in U$ so that $Fh \subset Fx$. Then for small $t > 0$ we have $x + th \in U$ and $F(x + th) = Fh$. It follows that $x + th \in M$ and $M \cap U \neq \emptyset$.

Finally, assume that $F(M)$ contains a dense G_δ completely metrizable set, D, say. Let d denote a complete metric generating the topology of D. For $n = 1, 2, \ldots$ we define sets

$$G_n = \left\{ x \in C(K) : \text{ there is an open set } Fx \subset U \subset K \right.$$

$$\left. \text{with } d\text{-diam} \ (\overline{U} \cap D) < \frac{1}{n} \right\}$$

and put $\Omega = \bigcap_{n=1}^{\infty} G_n$. Since F is upper semicontinuous, the sets G_n are open. Denote $H = \{x \in M : Fx \in D\}$; we can show, as in the previous paragraph, that H will still be dense in $C(K)$. Moreover, since $H \subset G_n$ for every $n \in \mathbb{N}$, the set Ω is dense in $C(K)$ too. Fix any $x \in \Omega$. It remains to show that Fx is a singleton. For each $n = 1, 2, \ldots$ we find an open set $Fx \subset U_n \subset K$ such that d-diam$(\overline{U_n} \cap D) < 1/n$. Note that $D \cap \bigcap_{i=1}^{n} \overline{U_i} \supset D \cap \bigcap_{i=1}^{n} U_i \neq \emptyset$ as D is dense in K and $\bigcap_{i=1}^{n} U_i \supset Fx \neq \emptyset$. It follows from the completeness of d that $D \cap \bigcap_{i=1}^{\infty} \overline{U_n}$ is a singleton, $\{k_0\}$, say. Assume there is $k \in Fx \backslash \{k_0\}$. Then there is an open set $k_0 \in W \subset K$ with $k \notin \overline{W}$. Thus for some $n \in \mathbb{N}$ we have $D \cap \overline{U_n} \subset W$ and so $k \in Fx \subset U_n \subset \overline{D \cap U_n} = D \cap \overline{U_n} \subset \overline{W}$, a contradiction. Hence $Fx = \{k_0\}$. $\square$

Theorem 2.2.3.

Let K be a compact space such that $C(K)$ is a Gâteaux differentiability space (a weak Asplund space). Then K is sequentially compact, and moreover, for every closed set $A \subset K$ the set Ω of G_δ points of A is dense in A (Ω contains a dense G_δ completely metrizable subset of A).

Proof. The sequential compactness follows from Theorem 2.1.2 because K continuously injects into $(C(K)^*, w^*)$. If $A = K$, the remaining assertions follow from Proposition 2.2.2. Now let A be any closed subset in K. We have two possible ways to proceed: Either we can use Proposition 2.2.2 when changing appropriately F_K and φ_K or we can observe that the mapping $x \mapsto x_{|A}$ maps $C(K)$ onto $C(A)$ [En, Theorem 2.1.8]; then we can use Theorems 2.1.3 and 2.1.5 and the just proved case $A = K$. $\square$

2.3. WEAK ASPLUNDNESS IS NOT A THREE-SPACE PROPERTY

We shall show a stronger fact:

Theorem 2.3.1.

There exist a Banach space D and its separable subspace Y (hence Y is WCG and so Asplund generated) such that the quotient D/Y is Asplund and WCG and yet D is not even weak Asplund.

Proof. Let D be the space of real-valued functions on $[0, 1]$ which are continuous from the left and have a finite limit from the right at every point $t \in [0, 1]$ and consider the supremum norm $\| \cdot \|$ on D. Then D is a Banach space. Perhaps we should check that for every $f \in D$ its norm $\|f\|$ is finite. Assume not. Then there are $t_n \in [0, 1]$ such that $|f(t_n)| \to +\infty$. Let t be a cluster point of $\{t_n\}$, say $t_{n_i} \to t$. Then surely $\lim \sup_i |f(t_{n_i})| < +\infty$, a contradiction.

For $t \in [0, 1)$ put $f_+(t) = \lim_{s \downarrow t} f(s)$. Consider the assignment $T : f \mapsto \{f_+(t) - f(t) : t \in [0, 1)\}$, $f \in D$. We claim that T is a linear continuous mapping from D onto $c_0([0, 1))$. Clearly T is linear and

$$\sup \{|f_+(t) - f(t)| : t \in [0, 1)\} \leq 2\|f\|, \qquad f \in D,$$

so that $\|T\| \leq 2$ and T is a continuous mapping from D to $\ell_\infty([0, 1))$. Fix $f \in D$ and $\epsilon > 0$. Assume there are countably many different $t_n \in [0, 1)$ such that $|Tf(t_n)| > \epsilon$. Let t be a cluster point of $\{t_n\}$, say, for simplicity, $t_n \to t$. Assume first that $t_n < t$ for all $n \in \mathrm{IN}$ large enough. Then $f(t_n) \to f(t)$ as well as $f_+(t_n) \to f(t)$. Hence $|f_+(t_n) - f(t_n)| > \epsilon$ only for finitely many $n \in \mathrm{IN}$, a contradiction. Second, there is an infinite set $N \subset \mathrm{IN}$ such that $t_n > t$ for every $n \in N$. Thus for $n \in N$ and $n \to +\infty$ we have $f(t_n) \to f_+(t)$, $f_+(t_n) \to f_+(t)$ and we get a similar nonsense. Therefore $Tf \in c_0([0, 1))$.

It remains to show the surjectivity of T. So take $x = \{x_t : t \in [0, 1)\}$ in the unit ball of $c_0([0, 1))$. Let $\{t_n\}$ be an enumeration of all $t \in [0, 1)$ for which $x_t \neq 0$. Fix $m \in \mathrm{IN} \cup \{0\}$. We put

$$A_m = \left\{ t \in [0, 1) : t = t_n \text{ for some } n \in \mathrm{IN} \text{ and } |x_t| \in (2^{-m-1}, 2^{-m}] \right\}$$

and let $s_1 < s_2 < \cdots < s_k$ be an increasing sequence such that $\{s_1, \ldots, s_k\} = A_m$. Define then $f_m : [0, 1] \to \mathrm{IR}$ as

$$f_m(t) = \begin{cases} 0 & \text{if } t \in [0, s_1], \\ x_{s_i} \dfrac{s_{i+1} - t}{s_{i+1} - s_i} & \text{if } t \in (s_i, s_{i+1}], \quad i = 1, 2, \ldots, k - 1, \\ x_{s_k} & \text{if } t \in (s_k, 1). \end{cases}$$

Then, putting $f = \sum_{m=0}^{\infty} f_m$, we have $f \in D$, $Tf = x$ and the surjectivity of T is proved.

Denote $Y = T^{-1}(0)$. It is easy to check that $Y = C([0,1])$. Thus Y is separable by Weierstrass's theorem [Ru2, Theorem 15.26] or according to [En, Corollary 4.2.18]. Since T is surjective, Banach's open mapping theorem guarantees that the quotient D/Y is isomorphic with $c_0([0,1))$ and this space is Asplund and WCG.

We shall show that D can be understood as a $C(K)$ space for some compact K. Put $K = [0,1] \cup \{t^+ : t \in [0,1]\}$. A topology on K is defined as follows. The points 0 and 1^+ are isolated. A neighborhood basis for $t \in (0,1]$ consists of the sets $(s,t] \cup \{r^+ : r \in (s,t)\}$, where $s \in [0,t)$. If $t \in [0,1)$, a neighborhood basis for t^+ is formed by the sets $(t,s) \cup \{r^+ : r \in [t,s)\}$, where $s \in (t,1]$. Such a K is called in the literature a *double-arrow* space. Let $\pi : K \rightarrow [0,1]$ be the mapping that assigns t to t as well as to t^+. Clearly π is continuous.

We claim *K is a compact space*. Let $\mathcal{U}$ be an open cover of K. For every $t \in [0,1]$ we find basic neighborhoods B_t and B_{t^+} of t and of t^+, respectively, and U_t, $U_{t^+} \in \mathcal{U}$, such that $B_t \subset U_t$, $B_{t^+} \subset U_{t^+}$ and put $\Omega_t = \pi(B_t \cup B_{t^+})$; note that this set contains t and is open in $[0,1]$. Hence there are $t_1, \ldots, t_n \in [0,1]$ such that $[0,1] = \Omega_{t_1} \cup \cdots \cup \Omega_{t_n}$. Fix now any $t \in [0,1]$. Then $t \in \Omega_{t_i}$ for some i. If $t \leq t_i$, then $t \in B_{t_i} (\subset U_{t_i})$. If $t > t_i$, then $t \in B_{t_i^+} (\subset U_{t_i^+})$. Similarly, $t < t_i$ implies $t^+ \in U_{t_i}$ and $t \geq t_i$ gives $t^+ \in U_{t_i^+}$. Therefore $K = U_{t_1} \cup U_{t_1^+} \cup \cdots \cup U_{t_n} \cup U_{t_n^+}$ and the compactness of K is proved.

For $f \in D$ we define $Sf : K \rightarrow \mathbb{R}$ by

$$Sf(t) = f(t) \qquad \text{and} \qquad Sf(t^+) = f_+(t), \qquad t \in [0,1].$$

It is easy to verify that S is a linear isometry from D into $C(K)$. Let us check that S is surjective. Take $g \in C(K)$. We define $f : [0,1] \rightarrow \mathbb{R}$ as $f(t) = g(t)$, $t \in [0,1]$. If $s \uparrow t$, then $s \rightarrow t$ in the topology of K and so $f(s) = g(s) \rightarrow g(t) = f(t)$. If $s \downarrow t$, then $s \rightarrow t^+$ in K and hence $f(s) = g(s) \rightarrow g(t^+)$. Thus $f \in D$ and we can easily check that $Sf = g$. Therefore D is isomorphic to $C(K)$.

If D were weak Asplund, then, according to Theorem 2.2.3, K would contain a dense G_δ completely metrizable space, M, say. In what follows we shall show that this is impossible. First we check that the set M is uncountable. Indeed, since K is compact, it is a Baire space [Ku, § 41.II, Theorem 6] and so is $K \setminus \{0, 1^+\}$. This space has no isolated points and $M \setminus \{0, 1^+\}$ is a residual set in it. Therefore the set M cannot be countable.

Now there are two possibilities. Either $(0,1) \cap M$ is uncountable or it is not. Assume the first case occurs. From the definition of K it is not difficult to check that any subspace of K is separable; in particular, so is $(0,1) \cap M$. Since M is metrizable, it follows that the space $(0,1) \cap M$ has a countable basis $\mathcal{B}$ for its topology, say $\mathcal{B} = \{B_n \cap (0,1) \cap M : n = 1,2,\ldots\}$, where B_n are open sets in K. Consider now any two distinct points $t_1, t_2 \in (0,1) \cap M$, say $t_1 < t_2$. Since $(0,t_i] \cap M$ is a neighborhood of t_i in $(0,1) \cap M$, there is $n_i \in \mathbb{N}$ such that $t_i \in B_{n_i} \cap (0,1) \cap M \subset (0,t_i] \cap M$, $i = 1,2$. We observe that $n_1 \neq n_2$. Indeed, if

$n_1 = n_2$, then $t_2 \in B_{n_2} \cap (0,1) \cap M \subset (0, t_1] \cap M$, a contradiction. In this way, we can construct a one-to-one assignment from $(0,1) \cap M$ into $\mathcal{B}$. However, this is impossible since $\mathcal{B}$ is countable and $(0,1) \cap M$ is not. In the second case, when $(0,1) \cap M$ is countable, the set $\{t^+ : t \in (0,1)\} \cap M$ is uncountable and we may proceed in a similar manner. Therefore the assumption that K contains a dense G_δ completely metrizable subspace proved to be false. Theorem 2.2.3 then says that $C(K)$, and hence D, is not a weak Asplund space. $\qquad \square$

2.4. NOTES AND REMARKS

Theorem 2.1.2 is from [St3]. That the converse statement is not true was proved by J. Bourgain; see [HS]. He constructed a Banach space $c_0 \subset X \subset \ell_\infty$ such that (B_{X^*}, w^*) is not sequentially compact while $(B_{(X/c_0)^*}, w^*)$ is. Assume that X/c_0 is a Gâteaux differentiability space. Then "weak* G_δ extreme points exist" in $(X/c_0)^*$. This property is a three-space property; so weak* G_δ extreme points exist in X^*. From this we can conclude that (B_{X^*}, w^*) is sequentially compact, a contradiction. See [LP] for details. Theorem 2.1.5 was already known to Asplund [As]. Another way to prove that $T(\Omega)$ is residual is by playing a Banach–Mazur game; see Lemma 4.2.1.

The further explanation follows freely the joint paper of Čoban and Kenderov [ČK]; see also [St11]. Theorem 2.1.5 was first announced by Asplund [As]; see also [Ph, Theorem 4.24].

The Gâteaux differentiability spaces were considered here in order to show a certain parallel with weak Asplund spaces. Perhaps we should mention the following chain of equivalent statements for a Banach space:

 (i) V is a Gâteaux differentiability space;
 (ii) Every weak* compact convex set K in the dual V^* has a weak* exposed point $\xi \in K$, that is, there is $v \in V$ such that $\langle \xi, v \rangle = \sup \langle K, v \rangle$ and $\langle \eta, v \rangle < \langle \xi, v \rangle$ for every $\eta \in K$, $\eta \neq \xi$;
 (iii) Every weak* compact convex set in the dual V^* is equal to the weak* closed convex hull of its weak* exposed points;
 (iv) $V \times \mathbb{R}$ is a Gâteaux differentiability space.

See the paper by Larman and Phelps [LP] and [Ph, Proposition 6.5] for proofs.

However, it is unclear if $V \times \mathbb{R}$ is weak Asplund whenever V is. Here we should quote a recent negative solution to Banach's hyperplane problem due to Gowers [Gow1, Gow2, GM]: There exists a separable Banach space not isomorphic to any of its hyperplanes, even to any proper subspace.

Theorem 2.2.3 suggests the following widely open question: Which additional properties should K satisfy so that $C(K)$ is a weak Asplund space (a Gâteaux differentiability space), or at least so that K belongs to Stegall's class S? See Chapter 3 for the definition of this class.

The space $D = C(K)$ in Section 2.3 was studied by Talagrand [Ta3]. He constructed on D a Lipschitz convex function whose set of points of Gâteaux differentiability is dense but not residual. It is shown in [Ta5; DGZ2, Theorem VII.3.5] that D *does not admit an equivalent Gâteaux smooth norm*. This fact will also follow from Theorem 2.3.1 and Corollary 4.2.5. Since every point of the space K is easily seen to be G_δ, we can conclude from Proposition 2.2.2 that the set of points where the (Lipschitz convex) function $f \mapsto \sup f(K)$ is Gâteaux differentiable is dense but not residual. Some easy additional gymnastics also yields that the supremum norm on $C(K)$ has the same property. However, it is not known *if $C(K)$, with this K, is a Gâteaux differentiability space*. Actually *there is no known example of a Gâteaux differentiability space which is not weak Asplund*.

Assuming the continuum hypothesis, Argyros and Mercourakis constructed *an equivalent norm on $\ell_1([0, \omega_1])$ whose set of points of Gâteaux differentiability is dense but not residual* [AM, Theorem 2.11]. Note that $\ell_1([0, \omega_1])$ is not a Gâteaux differentiability space since its canonical norm is nowhere Gâteaux differentiable.

We conclude by a necessary condition for weak Asplundness, even for Gâteaux differentiability spaces in one subclass of Banach spaces [AM] :

If V is a Gâteaux differentiability space and has an unconditional basis, then it is weakly Lindelöf determined and so (B_{V^*}, w^*) is a Corson compact (see Definition 7.2.6).

Proof. Let Γ be an unconditional basis in V, that is, Γ is a subset V with $\overline{\mathrm{sp}}\,\Gamma = V$ and there is $c > 0$ such that

$$\left\| \sum_{\gamma \in F} \epsilon_\gamma \alpha_\gamma \gamma \right\| \le c \left\| \sum_{\gamma \in F} \alpha_\gamma \gamma \right\|$$

whenever F is a finite set in Γ, $\epsilon_\gamma \in \{-1, 1\}$, $\alpha_\gamma \in \mathbb{R}$, $\gamma \in F$. We replace every $\gamma \in \Gamma$ by $\gamma/\|\gamma\|$; this new Γ will also be an unconditional basis in V. Assume that V is a Gâteaux differentiability space. We define $T : V^* \to \mathbb{R}^\Gamma$ by

$$Tv^*(\gamma) = \langle v^*, \gamma \rangle, \qquad v^* \in V^*, \qquad \gamma \in \Gamma.$$

Since T is weak*-to-pointwise continuous, we shall be done once we show that the set $\{\gamma \in \Gamma : \langle v^*, \gamma \rangle \ne 0\}$ is at most countable for every $v^* \in V^*$. Assume this is not the case. Then there are $v^* \in V^*$ and $\epsilon > 0$ such that the set $\Delta := \{\gamma \in \Gamma : |\langle v^*, \gamma \rangle| > \epsilon\}$ is uncountable. Take any nonempty finite set $F \subset \Delta$ and any $\alpha_\gamma \in \mathbb{R}$, $\gamma \in F$. For each $\gamma \in \Gamma$ we find $\epsilon_\gamma \in \{-1, 1\}$ such that $|\alpha_\gamma||\langle v^*, \gamma \rangle| = \langle v^*, \epsilon_\gamma \alpha_\gamma \gamma \rangle$. Then

$$\sum_{\gamma \in F} |\alpha_\gamma| \epsilon \leq \sum_{\gamma \in F} |\alpha_\gamma| |\langle v^*, \gamma \rangle| = \left\langle v^*, \sum_{\gamma \in F} \epsilon_\gamma \alpha_\gamma \gamma \right\rangle$$

$$\leq \|v^*\| \left\| \sum_{\gamma \in F} \epsilon_\gamma \alpha_\gamma \gamma \right\| \leq \|v^*\| c \left\| \sum_{\gamma \in F} \alpha_\gamma \gamma \right\| \leq \|v^*\| c \sum_{\gamma \in F} |\alpha_\gamma|.$$

It follows that $\ell_1(\Delta)$ is isomorphic to the subspace $Y := \overline{\mathrm{sp}}\{\gamma : \gamma \in \Delta\}$ of V. Now, since $\{\gamma : \gamma \in \Gamma\}$ is an unconditional basis for V, Y is complemented in V. Therefore the unit ball $\left(B_{\ell_\infty(\Delta)}, w^* \right)$ (homeomorphic to $[-1, 1]^\Delta$) can be found, up to a homeomorphism, in (V^*, w^*). Then, using the proof of Theorem 2.1.2, we can find a G_δ point of $[-1, 1]^\Delta$, say f. Let G_n be open sets in $[-1, 1]^\Delta$ such that $\{f\} = \bigcap_{n=1}^\infty G_n$. We find finite sets $F_n \subset \Delta$ and $\delta_n > 0$ such that

$$\{g \in [-1, 1]^\Delta : |f(\gamma) - g(\gamma)| < \delta_n, \ \gamma \in F_n\} \subset G_n.$$

Take $\gamma_0 \in \Delta \setminus \bigcup_{n=1}^\infty F_n$; it exists since Δ is uncountable. If we define $g \in [-1, 1]^\Delta$ by $g(\gamma) = f(\gamma)$ for $\gamma \in \Gamma \setminus \{\gamma_0\}$ and $g(\gamma_0) \neq f(\gamma_0)$, we get that $g \in \bigcap_{n=1}^\infty G_n$, a contradiction.   □

Finally, one comment concerning the definition of weak Asplund space. It is not difficult to see that a convex continuous function on a separable Banach space is Gâteaux differentiable exactly at the points of a dense G_δ set. In nonseparable Banach spaces this may not be the case. In fact, Holický, Šmídek, and Zajíček showed quite recently that *the set of Gâteaux differentiability of a convex continuous function defined* on a Banach space from quite a large class, in particular, *on a nonseparable Hilbert space, may not be G_δ or even Borel* [HŠZ].

Chapter Three

Stegall's Classes

There exist several permanence properties of the class of Asplund spaces: Asplundness is conserved after going to subspaces, quotients, finite products, and so on; see Theorem 1.1.2. By contrast, the situation in the weak Asplund case is quite unsatisfactory. No equivalent characterization is known and almost nothing can be said about the behavior of this class under standard Banach space operations. In particular, it is not clear whether weak Asplundness is inherited by uncomplemented subspaces or whether $V \times \mathbb{R}$ is weak Asplund when V is.

The aim of this chapter is to find a large (in fact the largest known) subclass of the weak Asplund spaces which is stable under standard Banach space operations and contains all the Asplund and WCG spaces. In the first section, we define a class S of topological spaces. This class is shown to enjoy several permanence properties. In the second section, a Banach space counterpart of S is considered: We define the class $\tilde{S}$ of those Banach spaces whose dual with the weak* topology belongs to S. Of course, we show that $\tilde{S}$ lies between the classes of Asplund spaces and weak Asplund spaces. The class $\tilde{S}$ also behaves very well; in particular, it contains the class $\mathcal{A}$ of all Banach spaces isomorphic to a subspace of an Asplund generated space. In this way we reprove Theorem 1.3.3.

3.1. STEGALL'S CLASS S OF TOPOLOGICAL SPACES

Let Z, X be two topological spaces and $F : Z \to 2^X$ be a multivalued mapping. We say that F is *upper semicontinuous* if for every open set $W \subset X$ the set $\{z \in Z : Fz \subset W\}$ is open, or equivalently, if for every closed set $C \subset X$ the set $\{z \in Z : Fz \cap C \neq \emptyset\}$ is closed. The mapping F is called *usco* if it is upper semicontinuous, and moreover, Fz is a nonempty compact set for every $z \in Z$. We say that F is a *minimal usco* mapping if it is usco, and moreover, $F = G$ whenever $G : Z \to 2^X$ is a usco mapping such that $G \subset F$. It easily follows from Zorn's lemma that every usco mapping contains at least one minimal usco mapping. The usco mappings, especially the minimal usco mappings, have several powerful, rather surprising properties. We collect them in the next two lemmas.

Lemma 3.1.1.

Let $F : Z \to 2^X$ be a mapping between topological spaces Z, X and assume that $Fz \neq \emptyset$ for each $z \in Z$. Then the following assertions are equivalent:

   (i) The mapping F is usco;

   (ii) If $\{(z_\tau, x_\tau)\} \subset F$ is a net such that $z_\tau \to z \in Z$, then a subnet of $\{x_\tau\}$ converges to an element of Fz;

   (iii) F is a closed set in $Z \times X$ and there exists a usco mapping $G : Z \to 2^X$ such that $F \subset G$.

Proof. Let (i) hold and assume that (ii) is false, that is, there is a net $\{(z_\tau, x_\tau)\} \subset F$ with $z_\tau \to z$ and such that no subnet of $\{x_\tau\}$ converges to an element of Fz. Then for every $x \in Fz$ there is an open set $x \in U_x \subset X$ such that $x_\tau \notin U_x$ for τ larger than some τ_x. Since Fz is compact, there are $x_1, \ldots, x_n \in Fz$ such that $Fz \subset U_{x_1} \cup \cdots \cup U_{x_n} =: U$. Then $x_\tau \notin U$ for all τ larger than $\tau_{x_1}, \ldots, \tau_{x_n}$, and this contradicts the upper semicontinuity of F at z. Therefore (ii) holds.

Now let (ii) be satisfied and fix $z \in Z$. Then, clearly, Fz is a compact set. If F were not upper semicontinuous at z, there would be an open set $Fz \subset U \subset X$ and a net $\{(z_\tau, x_\tau)\} \subset F$ with $z_\tau \to z$ and $x_\tau \notin U$. By (ii), a subnet $\{x_{\tau_i}\}$ converges to an $x \in Fz$. However, $x \notin U \ (\supset Fz)$, a contradiction. Hence (i) holds.

That (i) and (ii) imply (iii) is trivial.

Finally assume that (iii) is satisfied and let $\{(z_\tau, x_\tau)\}$ be a net in F. Then it is also a net in G. Hence there exists a subnet of $\{x_\tau\}$ converging to an element $x \in Gz$. Now $(z, x) \in F$ since F is closed. Thus F satisfies (ii). $\square$

Lemma 3.1.2.

Let $F : Z \to 2^X$ be a usco mapping between topological spaces Z and X. Then the following assertions are equivalent:

   (i) F is a minimal usco mapping;

   (ii) If $U \subset Z$ is an open set and $C \subset X$ is a closed set such that $Fz \cap C \neq \emptyset$ for each $z \in U$, then $F(U) \subset C$;

   (iii) If $F(U) \cap W \neq \emptyset$ for some open set $U \subset Z$ and some open set $W \subset X$, then there exists a nonempty open set $\Omega \subset U$ such that $F(\Omega) \subset W$;

   (iv) If $g : X \to Y$ is a continuous (single-valued) mapping from X to a topological space Y, then the composition $g \circ F : Z \to 2^Y$ is a minimal usco mapping.

Proof. (i)$\Rightarrow$(ii). Let U and C be as in (ii) and define $H : Z \to 2^X$ as

$$
Hz = \begin{cases} Fz \cap C & \text{if } z \in U, \\ Fz & \text{if } z \in Z \backslash U. \end{cases}
$$

Then for every $z \in Z$ the set Hz is nonempty and $H \subset F$. Take a net $\{(z_\tau, x_\tau)\}$ in H converging to $(z, x) \in Z \times X$. Then $x \in Fz$ by Lemma 3.1.1(ii), and $x \in C$ if $z \in U$. Hence $(z, x) \in H$ and Lemma 3.1.1(iii) guarantees that H is usco. Now (i) implies that $H = F$. Therefore, in particular, $Fz \subset C$ for every $z \in U$.

(ii)$\Rightarrow$(iii). Let U and W be as in (iii). If for every $z \in U$ the set $Fz \backslash W$ were nonempty, then we would have, by (ii), $F(U) \subset X \backslash W$, a contradiction. Hence, there is $z \in U$ such that $Fz \subset W$. Now the upper semicontinuity of F completes the proof.

(iii)$\Rightarrow$(i). Assume that $F_0 : Z \to 2^X$ is a minimal usco mapping with $F_0 \subset F$ and suppose $F_0 z_0 \neq F z_0$ for some $z_0 \in Z$. From the compactness of $F_0 z_0$ we find an open set $W \subset X$ such that $Fz_0 \cap W \neq \emptyset$ and $F_0 z_0 \cap \overline{W} = \emptyset$. The upper semicontinuity of F_0 guarantees that $F_0(U) \cap \overline{W} = \emptyset$ for some open set $z_0 \in U \subset Z$; then also $F(U) \cap W \neq \emptyset$. Hence, by (iii), $F(\Omega) \subset W$ for some nonempty open set $\Omega \subset U$. In particular, $F_0(\Omega) \subset W$, a contradiction.

(i)$\Rightarrow$(iv). Let $g : X \to Y$ be a continuous mapping. Then $g(Fz)$ will be a nonempty and compact set for every $z \in Z$. The composition $g \circ F$ is upper semicontinuous since for any closed set $C \subset Y$ the set

$$
\{z \in Z : (g \circ F)z \cap C \neq \emptyset\} = \{z \in Z : Fz \cap g^{-1}(C) \neq \emptyset\}
$$

is closed. The minimality of $g \circ F$ will be proved with the help of (iii). So consider open sets $U \subset Z$ and $W \subset Y$ such that $g \circ F(U) \cap W \neq \emptyset$. Then $F(U) \cap g^{-1}(W) \neq \emptyset$. Here $g^{-1}(W)$ is an open set and F is a minimal usco mapping. Hence there is a nonempty open set $\Omega \subset X$ such that $F(\Omega) \subset g^{-1}(W)$. Then $g \circ F(\Omega) \subset W$ and so $g \circ F$ must be minimal.

That (iv)$\Rightarrow$(i) is trivial. $\square$

Now we are ready to introduce the class S.

Definition 3.1.3.   The class S is the family of all completely regular spaces X such that whenever Z is a Baire space and $F : Z \to 2^X$ is a minimal usco mapping, there exists a residual subset Ω in Z such that Fz is a singleton, that is, it consists of one point, for every $z \in \Omega$.

Proposition 3.1.4.

Every metric space belongs to the class $\mathcal{S}$.

Proof. Let (X, ρ) be a metric space, Z be a Baire space, and $F: Z \to 2^X$ be a minimal usco mapping. For $n = 1, 2, \ldots$ define

$$G_n = \left\{ z \in Z : \text{ there is an open set } U \ni z \text{ with diameter of } F(U) \text{ less than } \frac{1}{n} \right\}.$$

Clearly, the sets G_n are open; if z belongs to all of them, then Fz is surely a singleton. It remains to prove that the sets G_n are dense. So fix any $n \in \mathbb{N}$ and let $\emptyset \neq G \subset Z$ be any open set. Let W be an open ball in X with radius $1/(3n)$ such that $F(G) \cap W \neq \emptyset$. By Lemma 3.1.2(iii), there is an open set $\emptyset \neq U \subset G$ such that $F(U) \subset W$. Hence $G_n \cap G \supset U \neq \emptyset$ and the density of G_n is proved. $\qquad\square$

More interesting examples of spaces belonging to $\mathcal{S}$ will be dealt with in chapters 4, 5, and 7. Now let us pass to the promised permanence properties.

Theorem 3.1.5.

Assume that all the spaces below are completely regular.
- *(i) If $X \in \mathcal{S}$ and $g: X \to Y$ is a surjective perfect mapping, that is, g is continuous and maps closed sets onto closed sets and $g^{-1}(y)$ is a compact set for each $y \in Y$, then $Y \in \mathcal{S}$; in particular, a continuous image of a compact from $\mathcal{S}$ belongs to $\mathcal{S}$.*
- *(ii) If $Y \in \mathcal{S}$ and $g: X \to Y$ is continuous and one-to-one, then $X \in \mathcal{S}$; in particular, every subspace of Y belongs to $\mathcal{S}$.*
- *(iii) If $X_n \in \mathcal{S}$, $n = 1, 2, \ldots$, then $\prod_{n=1}^{\infty} X_n \in \mathcal{S}$.*
- *(iv) If $X = \bigcup_{n=1}^{\infty} X_n$ and each X_n is a closed subset of X and belongs to $\mathcal{S}$, then $X \in \mathcal{S}$.*

Proof. (i) Let X, Y, g be as in the premise. Let Z be a Baire space and $F: Z \to 2^Y$ be minimal usco. Then the mapping $g^{-1} \circ F: Z \to 2^X$ assigning to each $z \in Z$ the set $g^{-1}(Fz)$ is upper semicontinuous. Indeed, if $C \subset X$ is a closed set, then, by the perfectness of g, the set $g(C)$ is closed and so the set

$$\{z \in Z : (g^{-1} \circ F)z \cap C \neq \emptyset\} = \{z \in Z : Fz \cap g(C) \neq \emptyset\}$$

is closed owing to the upper semicontinuity of F. The mapping $g^{-1} \circ F$ is even usco since $g^{-1}(K)$ *is compact for any compact set* $K \subset Y$.

In order to prove the proclaimed compactness of $g^{-1}(K)$, fix a compact set $K \subset Y$ and consider a system $\{C_\gamma\}_{\gamma \in \Gamma}$ of nonempty closed subsets of $g^{-1}(K)$ which is closed under finite intersections. We have to show that $\bigcap_{\gamma \in \Gamma} C_\gamma \neq \emptyset$

[En, Theorem 3.1.1]. By perfectness of g, all the sets $g(C_\gamma)$ are closed. Further, as K is compact, $\bigcap_{\gamma \in \Gamma} g(C_\gamma) \neq \emptyset$. Take k in this intersection. Thus $C_\gamma \cap g^{-1}(k) \neq \emptyset$ for each $\gamma \in \Gamma$. Now $g^{-1}(k)$ is compact as g is perfect. Hence $\bigcap_{\gamma \in \Gamma} (C_\gamma \cap g^{-1}(k)) \neq \emptyset$, and, a fortiori, $\bigcap_{\gamma \in \Gamma} C_\gamma \neq \emptyset$. This finishes the proof that $g^{-1}(K)$ is compact.

Let us recall that we have so far shown that $g^{-1} \circ F$ is usco. Let $H : Z \to 2^X$ be a minimal usco mapping such that $H \subset g^{-1} \circ F$. Since $X \in S$, there is a residual set $\Omega \subset Z$ such that Hz is a singleton for every $z \in \Omega$. Now we remark that $g \circ H$ is also usco (See the proof of Lemma 3.1.2(iv).) and that $g \circ H \subset F$. But F is minimal, so $g(Hz) = Fz$ for each $z \in Z$. Therefore Fz is a singleton for every $z \in \Omega$. We have thus shown that $Y \in S$.

(ii) Let X, Y, g be as in the assumptions and let $F : Z \to 2^X$ be a minimal usco mapping where Z is a Baire space. According to Lemma 3.1.2(iv), $g \circ F : Z \to 2^Y$ is minimal usco, too. Hence, $g(Fz)$ is a singleton for all z from a residual subset of Z, and since g is injective, we are done.

(iii) Let $X_n \in S$, $n = 1, 2, \ldots$, let Z be a Baire space, and consider a minimal usco mapping F from Z into $X := \prod_{n=1}^{\infty} X_n$. Denote by π_n the canonical projection from X onto X_n. Then $\pi_n \circ F$ will be minimal usco mappings by Lemma 3.1.2(iv). As $X_n \in S$, there exist residual sets $\Omega_n \subset Z$ such that $\pi_n \circ F$ are single-valued at the points of Ω_n. Put $\Omega = \bigcap_{n=1}^{\infty} \Omega_n$. This set is still residual and, surely, Fz is a singleton for each $z \in \Omega$.

(iv) Let X_n, X be as in the premise, let Z be a Baire space, and let $F : Z \to 2^X$ be a minimal usco mapping. For $n = 1, 2, \ldots$ we define $Z_n = \{z \in Z : Fz \cap X_n \neq \emptyset\}$ and let U_n denote the interior of Z_n. Since Z is a Baire space and $\bigcup_{n=1}^{\infty} Z_n = Z$, the set $\bigcup_{n=1}^{\infty} U_n$ is dense in Z.

Let Ω denote the set of all $z \in Z$ for which Fz is a singleton and fix any $n \in \text{IN}$ such that $U_n \neq \emptyset$. We can easily verify that the restriction $F_{|U_n}$ of F to U_n is a minimal usco mapping which sends U_n into X_n; see Lemma 3.1.2. Now, as $X_n \in S$, the set $\Omega \cap U_n$ is residual in U_n. Now Proposition 1.3.2 guarantees that the set Ω is residual in Z. $\qquad\qquad\square$

Theorem 3.1.6.

Let K be a compact space lying in S. Then

(i) K contains a dense G_δ completely metrizable set and

(ii) K is sequentially compact, that is, every sequence in K has a convergent subsequence.

Proof. Let us define a multivalued mapping $F : C(K) \to 2^K$ by

$$F(f) = \{k \in K : f(k) = \sup f(K)\}, \qquad f \in C(K).$$

We already know from Lemma 2.2.1(i) that F is a usco mapping. It is in fact minimal usco. Indeed, assume that $F(U) \cap W \neq \emptyset$, where $U \subset C(K)$ and

$W \subset K$ are open sets. Pick $f \in U$ such that $F(f) \cap W \neq \emptyset$ and find $k_0 \in F(f) \cap W$. By [En, Theorem 1.5.10], we find $h \in C(K)$ such that $h(k_0) = \sup h(K)$ and $h(k) = 0$ for every $k \in K \setminus W$. Let $t > 0$ be so small that $f + th \in U$. Then $F(f + th) \subset W$ and the upper semicontinuity of F yields that $F(\Omega) \subset W$ for some nonempty open set $\Omega \subset U$. Therefore, F is a minimal usco mapping according to Lemma 3.1.2(iii).

Assertion (ii) is an immediate consequence of (i), Theorem 3.1.5(ii), and Lemma 2.1.1.   □

3.2. STEGALL'S CLASS $\widetilde{S}$ OF BANACH SPACES

Definition 3.2.1.   The class $\widetilde{S}$ consists of all those Banach spaces whose dual endowed with the weak* topology belongs to S.

Here we should recall that *every dual Banach space with the weak* topology is completely regular*. Indeed, let V^* be a dual Banach space, $\xi_0 \in V^*$, and let $\xi_0 \in W \subset V^*$ be a weak* open set. Then there are $v_1, \ldots, v_k \in V$ such that $\xi \in W$ whenever $\xi \in V^*$ and $|\langle \xi - \xi_0, v_i \rangle| < 1$, $i = 1, \ldots, k$. Define the function $f : V^* \to [0, 1]$ by

$$f(\xi) = \min\left\{1, \max\{|\langle \xi - \xi_0, v_1 \rangle|, \ldots, |\langle \xi - \xi_0, v_k \rangle|\}\right\}, \qquad \xi \in V^*.$$

Then $f(\xi_0) = 0$ and $f(\xi) = 1$ for all $\xi \in V^* \setminus W$, which shows that (V^*, w^*) is completely regular.

According to Theorem 3.1.5(ii), (iv), we know that $V \in \widetilde{S}$ if and only if the unit ball B_{V^*} of V^*, endowed with the weak* topology, belongs to S. Then, by Proposition 3.1.4, every separable Banach space is in $\widetilde{S}$.

Now the reader probably expects the next theorem.

Theorem 3.2.2.

The class $\widetilde{S}$ lies between the Asplund and the weak Asplund spaces, that is, every Asplund space belongs to $\widetilde{S}$ and every element of $\widetilde{S}$ is weak Asplund.

Proof.   Let V be an Asplund space. It means, by Theorem 1.1.1, that the dual V^* is weak* dentable, that is, that for every $\epsilon > 0$ and every bounded set $\emptyset \neq M \subset V^*$ there are $e \in V$ and $\alpha > 0$ such that the weak* slice

$$S(M, e, \alpha) = \{v^* \in M : \langle v^*, e \rangle > \sup\langle M, e \rangle - \alpha\}$$

has norm-diameter less than ϵ. Let Z be a Baire space and let $F: Z \to 2^{(B_{V^*}, w^*)}$ be a minimal usco mapping. For $n = 1, 2, \ldots$ we consider the sets

$$U_n = \Big\{ z \in Z : \quad \text{there is an open set } z \in \Omega \subset Z$$

$$\text{such that norm-diameter of } F(\Omega) < \frac{1}{n} \Big\}.$$

Clearly, each set U_n is open. We shall show that each U_n is dense in Z. So let $n \in \mathbb{N}$ be given and let $\emptyset \neq U \subset Z$ be any open set. As $F(U) \subset B_{V^*}$, there are, by Theorem 1.1.1, $e \in V$ and $\alpha > 0$ such that the weak* slice $S(F(U), e, \alpha)$ has norm-diameter less than $1/n$. Let us observe that we can write $S(F(U), e, \alpha) = F(U) \cap W$, where

$$W = \big\{ v^* \in B_{V^*} : \quad \langle v^*, e \rangle > \sup \langle F(U), e \rangle - \alpha \big\}.$$

By Lemma 3.1.2(iii), there is a nonempty open set $\Omega \subset U$ such that $F(\Omega) \subset W$ and so $\Omega \subset U_n \cap U$. Thus the density of U_n is proved. It should also be clear that for every $z \in \bigcap_{n=1}^{\infty} U_n$ the set Fz is a singleton. We have thus verified that (B_{V^*}, w^*) belongs to S and so $V \in \widetilde{S}$.

It remains to prove the second part of the theorem; let V be a Banach space with (V^*, w^*) in S. Consider a continuous convex function $f: V \to \mathbb{R}$. We recall that the subdifferential mapping $\partial f : V \to 2^{V^*}$ is defined by

$$\partial f(v) = \big\{ v^* \in V^* : \quad f(v + h) - f(v) \geq \langle v^*, h \rangle \quad \text{for all} \quad h \in V \big\}$$

and that ∂f is a usco mapping when V is endowed with the norm topology and V^* is endowed with the weak* topology; see [Ph, Proposition 2.5]. Let $F: V \to 2^{(V^*, w^*)}$ be a minimal usco mapping such that $F \subset \partial f$. By the definition of the class $\widetilde{S}$, there is a residual set Ω in V such that Fv is a singleton whenever $v \in \Omega$. Fix one $v \in \Omega$. Then for all $h \in V$ and all $t > 0$ we have, by monotonicity,

$$\langle Fv, h \rangle \leq \frac{1}{t} \big[f(v + th) - f(v) \big] \leq \langle \xi_t, h \rangle,$$

where ξ_t is any element of $F(v + th)$. Now we use the norm-to-weak* upper semicontinuity of F to conclude that

$$\langle Fv, h \rangle = \lim_{t \downarrow 0} \frac{1}{t} \big[f(v + th) - f(v) \big]$$

for all $h \in V$. This means that f is Gâteaux differentiable at v with the derivative Fv and the weak Asplundness of V is verified.   □

Now the permanence properties of class $\widetilde{S}$ follow.

Theorem 3.2.3.

(i) If $V \in \widetilde{S}$ and $T : V \to Y$ is a bounded linear mapping from V into a Banach space Y with dense range, then $Y \in \widetilde{S}$; in particular, if $V \in \widetilde{S}$, then every quotient of V is in $\widetilde{S}$.

(ii) If $Y \in \widetilde{S}$ and $T : V \to Y$ is linear bounded and such that its second conjugate T^{**} is one-to-one, that is, $T^*(Y^*)$ is dense in V^*, then $V \in \widetilde{S}$; in particular, the class $\widetilde{S}$ is closed under going to subspaces.

(iii) If $V_n \in \widetilde{S}$, $n = 1, 2, \ldots$, then $\left(\sum_{n=1}^{\infty} V_n \right)_{\ell_1} \in \widetilde{S}$.

(iv) If V_n, $n = 1, 2, \ldots$, are subspaces of a Banach space V such that $\mathrm{sp} \left(\bigcup_{n=1}^{\infty} V_n \right)$ is dense in V and each $V_n \in \widetilde{S}$, then $V \in \widetilde{S}$.

(v) If Γ is an arbitrary nonempty set and $V_\gamma \in \widetilde{S}$ for each $\gamma \in \Gamma$, then $\left(\sum_{\gamma \in \Gamma} V_\gamma \right)_{c_0} \in \widetilde{S}$ and $\left(\sum_{\gamma \in \Gamma} V_\gamma \right)_{\ell_p} \in \widetilde{S}$ for all $p \in (1, +\infty)$.

(vi) If an Asplund space V is a subspace of a Banach space Y and the quotient Y/V is in $\widetilde{S}$, then $Y \in \widetilde{S}$.

Proof. (i) Let V, Y, and T be as in the premise. Then $T^* : (Y^*, w^*) \to (V^*, w^*)$ is one-to-one and continuous. Hence, by Theorem 3.1.5(ii), $Y \in \widetilde{S}$.

(iii) Let all V_n, $n = 1, 2, \ldots$, be in $\widetilde{S}$. We recall that $V := \left(\sum_{n=1}^{\infty} V_n \right)_{\ell_1}$ means the Banach space of those sequences $\{v_n\}$ from $\prod_{n=1}^{\infty} V_n$ for which $\sum_{n=1}^{\infty} \|v_n\| < +\infty$. It is a routine matter to verify that (B_{V^*}, w^*) is homeomorphic with $\prod_{n=1}^{\infty} (B_{V_n^*}, w^*)$. Since the last space is in S by Theorem 3.1.5(iii), we conclude that $V \in \widetilde{S}$.

(iv) Let V_n and V be as in the premise. Then $\left(\sum_{n=1}^{\infty} V_n \right)_{\ell_1} \in \widetilde{S}$ by (iii). Moreover, the mapping $\{v_n\} \mapsto \sum_{n=1}^{\infty} v_n$ is linear continuous and maps $\left(\sum_{n=1}^{\infty} V_n \right)_{\ell_1}$ onto a dense set in V. Hence (ii) applies.

The proofs of (ii), (v), and (vi) require more work. We shall need the following:

Proposition 3.2.4.

Let Z be a Baire space, V be a Banach space, and $F : Z \to 2^{(V^*, w^*)}$ be a minimal usco mapping. For $n = 1, 2, \ldots$ define a family $\mathcal{U}_n$ consisting of all open sets $U \subset Z$ for which there exists a weak* closed set $C_U \subset V^*$ such that $(C_U, w^*) \in S$ and $F(U) \subset C_U + (1/n)B_{V^*}$. Assume that for each $n = 1, 2, \ldots$ the union of the sets from the family $\mathcal{U}_n$ is dense in Z. Then F is single-valued at the points of a residual subset of Z.

Proof. Fix any $n \in \mathbb{N}$ and any $U \in \mathcal{U}_n$. Let C_U be the weak* closed set corresponding to U. Define a mapping $H : U \to 2^{(C_U, w^*)}$ by

$$Hz = \left(Fz + \frac{1}{n} B_{V^*} \right) \cap C_U, \qquad z \in U.$$

Clearly, $Hz \neq \emptyset$ for every $z \in U$. We shall show that H is usco. Let $\{(z_\tau, \xi_\tau)\} \subset H$ be a net such that $z_\tau \to z \in Z$. For each τ we find $\eta_\tau \in (1/n) B_{V^*}$ such that $\xi_\tau - \eta_\tau \in F z_\tau$. By Lemma 3.1.1, there is a subnet $\{\xi_{\tau_i} - \eta_{\tau_i}\}$ weak* converging to some $\zeta \in Fz$. Owing to the weak* compactness of $(1/n) B_{V^*}$ we may also assume that $\{\eta_{\tau_i}\}$ weak* converges to some $\eta \in (1/n) B_{V^*}$. Thus $\xi_{\tau_i} \to \zeta + \eta$ weak*. Now, as C_U was weak* closed, we can conclude that $\zeta + \eta \in Hz$. Finally, Lemma 3.1.1 guarantees that H is usco.

Let $H_0 : U \to 2^{(C_U, w^*)}$ be a minimal usco mapping such that $H_0 \subset H$ and find any (single-valued) mapping $f_U^n : U \to C_U$ such that $f_U^n(z) \in H_0(z)$ for each $z \in U$. Then, clearly, for every $z \in U$

$$\operatorname{dist}\left(f_U^n(z), Fz\right) := \inf\{ \| f_U^n(z) - \xi \| : \xi \in Fz \} \leq \frac{1}{n}.$$

Now, since $(C_U, w^*) \in \mathcal{S}$, it follows that f_U^n is continuous at the points of a residual subset Ω_U^n of U.

Next put

$$\Omega_n = \bigcup \{ \Omega_U^n : U \in \mathcal{U}_n \}, \qquad n = 1, 2, \dots .$$

Since the union of the sets from $\mathcal{U}_n$ is dense in Z, we know by Proposition 1.3.2 that the sets Ω_n are residual in Z. Finally put $\Omega = \bigcap_{n=1}^{\infty} \Omega_n$. This set is also residual in Z. It remains to show that Ω is the set we have been looking for. Fix any z_0 in Ω. We have to prove that Fz_0 is a singleton. Let W be an arbitrary convex weak* closed weak* neighborhood of the origin in V^*. For each $n \in \mathbb{N}$ we find $U_n \in \mathcal{U}_n$ such that $z_0 \in U_n$. Let $\xi_0 \in V^*$ be a weak* cluster point of the sequence $\{ f_{U_n}^n(z_0) \}$; it exists since Fz_0 is weak* compact and $\operatorname{dist}\left(f_{U_n}^n(z_0), Fz_0\right) \to 0$ for $n \to \infty$. Choose $n \in \mathbb{N}$ so large that $(3/n) B_{V^*} \subset W$. There exists $m \in \mathbb{N}$, $m \geq n$, such that $f_{U_m}^m(z_0) \in \xi_0 + \frac{1}{3} W$. Since $f_{U_m}^m$ is continuous at z_0, there is an open set $z_0 \in G \subset U_m$ such that $f_{U_m}^m(G) \subset \xi_0 + \frac{2}{3} W$. Thus

$$f_{U_m}^m(G) + \frac{1}{m} B_{V^*} \subset \xi_0 + \frac{2}{3} W + \frac{1}{m} B_{V^*} \subset \xi_0 + W$$

and the definition of $f_{U_m}^m$ ensures that $Fz \cap (\xi_0 + W) \neq \emptyset$ for each $z \in G$. But F is minimal usco and W is weak* closed. Hence by Lemma 3.1.2(ii) we have $F(G) \subset \xi_0 + W$. Then, in particular, $Fz_0 \subset \xi_0 + W$. Since W was arbitrary, we conclude that $Fz_0 = \{\xi_0\}$. $\qquad \square$

Proof of Theorem 3.2.3 (ii), (v), and (vi). (ii) Let V, Y, T be as in the premise, that is, $T : V \to Y$, $Y \in \tilde{\mathcal{S}}$, and $T^* Y^*$ is dense in V^*. We have to show that V belongs to $\tilde{\mathcal{S}}$. Let Z be a Baire space and let $F : Z \to 2^{(B_{V^*}, w^*)}$ be a minimal usco mapping. For $n = 1, 2, \ldots$ define

$$\mathcal{U}_n = \left\{ U \subset Z : \ U \text{ open, } F(U) \subset T^*(m B_{Y^*}) + \frac{1}{n} B_{V^*} \text{ for some } m > 0 \right\}.$$

Here $T^*(m B_{Y^*})$ is a weak* compact, and hence weak* closed, set which belongs to $\mathcal{S}$ by Theorem 3.1.5. Thus, according to Proposition 3.2.4, we shall be done when we show that the union of the sets from $\mathcal{U}_n$ is dense in Z for each n. So fix any $n \in \mathbb{N}$ and let G be any nonempty open subset of Z. Since $T^*(Y^*)$ is dense in V^*, it follows that $F(G) \subset \bigcup_{m=1}^{\infty} T^*(m B_{Y^*}) + (1/n) B_{V^*}$. Denote

$$M_m = \left\{ z \in G : \ Fz \cap \left(T^*(m B_{Y^*}) + \frac{1}{n} B_{V^*} \right) \neq \emptyset \right\}, \qquad m = 1, 2, \ldots \ .$$

These sets are closed in G and their union is equal to G. Hence, as G is a Baire space, there is $m \in \mathbb{N}$ such that M_m has a nonempty interior, U, say. Then, by Lemma 3.1.2(ii), we get $F(U) \subset T^*(m B_{Y^*}) + (1/n) B_{Y^*}$. This means that U belongs to $\mathcal{U}_n$. Therefore G intersects the union of $\mathcal{U}_n$. Now we can apply Proposition 3.2.4 to conclude the proof of (ii).

Note that, in the particular case, when V is a subspace of Y and $Y \in \tilde{\mathcal{S}}$, there is a simpler proof that $V \in \tilde{\mathcal{S}}$. In fact, then, $(Y^*, w^*) \in \mathcal{S}$ and hence $(B_{Y^*}, w^*) \in \mathcal{S}$. Now, the assignment $\xi \mapsto \xi_{|V}$ is a perfect mapping from (B_{Y^*}, w^*) onto (B_{V^*}, w^*). Thus, by Theorem 3.1.5(i), (iv), $(B_{V^*}, w^*) \in \mathcal{S}$, $(V^*, w^*) \in \mathcal{S}$; so $V \in \tilde{\mathcal{S}}$.

(v) Assume we have a nonempty set Γ and Banach spaces $(V_\gamma, \|\cdot\|_\gamma) \in \tilde{\mathcal{S}}$, $\gamma \in \Gamma$. Let V and Y_p, $1 < p < +\infty$, denote the c_0-sum and the ℓ_p-sum of the V_γ, $\gamma \in \Gamma$, respectively, that is,

$$V = \left\{ \{v_\gamma\} \in \prod_{\gamma \in \Gamma} V_\gamma : \ \{\|v_\gamma\|_\gamma\} \in c_0(\Gamma) \right\},$$

$$Y_p = \left\{ \{v_\gamma\} \in \prod_{\gamma \in \Gamma} V_\gamma : \ \{\|v_\gamma\|_\gamma\} \in \ell_p(\Gamma) \right\}.$$

Let us remark that once having proved that $V \in \tilde{\mathcal{S}}$, Theorem 3.2.3(ii) will yield that each Y_p, with $p \in (1, +\infty)$, belongs to $\tilde{\mathcal{S}}$. Indeed, fix any $1 < p < +\infty$ and consider the mapping $T : Y_p \to V$ defined by $T(\{v_\gamma\}) = \{v_\gamma\}$, $\{v_\gamma\} \in Y_p$. Then T is linear and $\|T\| \leq 1$. We have to show that $T^* V^*$ is dense in Y_p^*. Let η be any element of Y_p^*. For $\gamma \in \Gamma$ we put $\eta_\gamma(v_\gamma) = \langle \eta, \{0, \ldots, 0, v_\gamma, 0, \ldots\} \rangle$, $v_\gamma \in V_\gamma$. Here, of course, v_γ is at the γ-th position. Clearly, $\eta_\gamma \in V_\gamma^*$. By imitating a standard "calculation" of the dual to ℓ_p, we can verify that $\|\eta\| = \left(\sum_{\gamma \in \Gamma} \|\eta_\gamma\|_\gamma^q \right)^{1/q}$, where q satisfies the equation $1/p + 1/q = 1$; see, for instance, [Tay, Example 4.2.4, Section 4.32]. Let $\epsilon > 0$ be given. We find a finite set $A \subset \Gamma$ such that $\|\eta\|^q - \epsilon < \sum_{\gamma \in A} \|\eta_\gamma\|^q_\gamma$. Put $\xi(\{v_\delta\}) = \sum_{\gamma \in A} \langle \eta_\gamma, v_\gamma \rangle$, $\{v_\delta\} \in V$. Then $\xi \in V^*$, $T^*\xi \in Y^*_p$ and we can estimate

$$\|\eta - T^*\xi\|^q = \sum_{\gamma \in \Gamma \setminus A} \|\eta_\gamma\|^q_\gamma = \|\eta\|^q - \sum_{\gamma \in A} \|\eta_\gamma\|^q_\gamma < \epsilon.$$

This shows that T^*V^* is dense in Y^*_p and Theorem 3.2.3(ii) gives that $Y_p \in \tilde{S}$.

Now it remains to prove that $V \in \tilde{S}$. We shall again apply Proposition 3.2.4. Let F be a minimal usco mapping from a Baire space Z into (V^*, w^*). Define $\varphi : Z \to [0, +\infty)$ by

$$\varphi(z) = \min\{\|\xi\| : \xi \in Fz\}, \qquad z \in Z.$$

We can easily verify that this function is lower semicontinuous. Let us introduce the following notation. For $A \subset \Gamma$ we put

$$V^*_A = \Big\{ \xi \in V^* : \langle \xi, v \rangle = 0 \text{ whenever } v = \{v_\gamma\} \in V$$

$$\text{and } v_\gamma = 0 \text{ for all } \gamma \in \Gamma \setminus A \Big\}.$$

Clearly, if A is a finite set, the space (V^*_A, w^*) is homeomorphic with $\prod_{\gamma \in A} (V^*_\gamma, w^*)$. Hence, by Theorem 3.1.5(iii), $(V^*_A, w^*) \in S$.

For $n = 1, 2, \ldots$ we introduce the families

$$\mathcal{U}_n = \Big\{ U \subset Z : U \text{ open and } F(U) \subset V^*_A + \frac{1}{n}B_{V^*}$$

$$\text{for some finite set } A \subset \Gamma \Big\}.$$

We have to show that the union of the sets from $\mathcal{U}_n$ is dense in Z. Fix any nonempty open set Ω in Z. Since φ is lower semicontinuous, it is continuous at the points of a dense G_δ subset of Z [En, 1.7.14]; choose $z \in \Omega$ such that φ is continuous at z. Then there is an open set $z \in G \subset \Omega$ such that $\varphi(z') < \varphi(z) + (2n)^{-1}$ for each $z' \in G$. By the minimality of F, we get that $F(G) \subset \big(\varphi(z) + (2n)^{-1} \big) B_{V^*}$. We can also find $\xi \in Fz$ such that $\|\xi\| = \varphi(z)$. Further, there is a finite set $A \subset \Gamma$ such that $\|\xi\| - (2n)^{-1} < \sum_{\gamma \in A} \|\xi_\gamma\|_\gamma$; here $\xi_\gamma \in V^*_\gamma$ are defined by $\langle \xi_\gamma, v_\gamma \rangle = \langle \xi, \{0, \ldots, 0, v_\gamma, 0 \ldots\} \rangle$, $v_\gamma \in V_\gamma$, with v_γ at the γ-th position. Finally, for every $\gamma \in A$ we find $v_\gamma \in V_\gamma$, with $\|v_\gamma\|_\gamma = 1$, such that

$$\sum_{\gamma \in A} \langle \xi_\gamma, v_\gamma \rangle > \|\xi\| - (2n)^{-1}.$$

Now consider the (weak* open) set

$$W = \Big\{ \eta \in V^* : \sum_{\gamma \in A} \langle \eta_\gamma, v_\gamma \rangle > \|\xi\| - (2n)^{-1} \Big\}.$$

(Here η_γ is obtained from η as ξ_γ was from ξ.) Then $Fz \cap W$ contains ξ. Hence, by Lemma 3.1.2(iii), there is an open set $\emptyset \neq U \subset G$ so that $F(U) \subset W$. Take any $\eta \in F(U)$; then

$$\|\eta\| \leq \varphi(z) + (2n)^{-1} \quad \text{and} \quad \sum_{\gamma \in A} \langle \eta_\gamma, v_\gamma \rangle > \|\xi\| - (2n)^{-1} = \varphi(z) - (2n)^{-1}.$$

So, defining $\eta_A \in V_A^*$ by the formula $\langle \eta_A, \{u_\delta\} \rangle = \sum_{\gamma \in A} \langle \eta_\gamma, u_\gamma \rangle$, $\{u_\delta\} \in V$, we get

$$\|\eta - \eta_A\| = \sum_{\gamma \in \Gamma \setminus A} \|\eta_\gamma\|_\gamma = \|\eta\| - \sum_{\gamma \in A} \|\eta_\gamma\|_\gamma \leq \|\eta\| - \sum_{\gamma \in A} \langle \eta_\gamma, v_\gamma \rangle$$

$$\leq \varphi(z) + (2n)^{-1} - \left(\varphi(z) - (2n)^{-1} \right) = n^{-1}.$$

Hence

$$\eta = \eta_A + (\eta - \eta_A) \in V_A^* + \frac{1}{n} B_{V^*},$$

and this holds for every $\eta \in F(U)$. Thus $U \in \mathcal{U}_n$, which proves the density of the union of $\mathcal{U}_n$. Finally, recalling that $(V_A^*, w^*) \in \mathcal{S}$, we can conclude, using Proposition 3.2.4, that $V \in \widetilde{\mathcal{S}}$. This proves (v).

(vi) Assume that V is Asplund, that V is a subspace of a Banach space Y, and that Y/V belongs to $\widetilde{\mathcal{S}}$. We have to prove that $Y \in \widetilde{\mathcal{S}}$. Let $F : Z \to 2^{(Y^*, w^*)}$ be a minimal usco mapping, where Z is a Baire space. We may and do assume that $F(Z) \subset B_{Y^*}$. We shall again use Proposition 3.2.4. Denote by i the natural embedding of V into Y. Then, by Lemma 3.1.2(iv), we know that $i^* \circ F : Z \to 2^{(B_{V^*}, w^*)}$ is a minimal usco mapping as well. For $n = 1, 2, \ldots$ consider the families

$$\mathcal{U}_n = \left\{ U \subset Z : \ U \text{ is open and the diameter of } i^*(F(U)) \text{ is less than } \frac{1}{n} \right\}.$$

We shall show that the union of the sets from $\mathcal{U}_n$ is dense in Z for each $n \in \mathrm{IN}$. So fix $n \in \mathrm{IN}$ and let Ω be a nonempty open subset of Z. Since V^* is, by Theorem 1.1.1, weak* dentable, there are $e \in V$ and $\alpha > 0$ such that the weak* slice $S(i^*(F(\Omega)), e, \alpha)$ has diameter less than $1/n$. Hence there is a weak* open set $W \subset V^*$ such that the set $i^*(F(\Omega)) \cap W$ is nonempty and has diameter less than $1/n$. Now Lemma 3.1.2(iii) says that there is a nonempty open set $U \subset \Omega$ such that $i^*(F(U)) \subset W$. Hence the diameter of $i^*(F(U))$ is less than $1/n$. This shows that U belongs to $\mathcal{U}_n$ and so the density of the union of $\mathcal{U}_n$ is proved.

Next, fix any $U \in \mathcal{U}_n$. Thus there is $v^* \in V^*$ such that $i^*(F(U)) \subset v^* + (1/n)B_{V^*}$. Find $y^* \in Y^*$ such that $i^*(y^*) = v^*$; its existence is guaranteed by the Hahn–Banach theorem. Then $F(U) \subset y^* + (1/n)B_{Y^*} + V^\perp$, where $V^\perp = \{\eta \in Y^* : \ \langle \eta, v \rangle = 0 \text{ for all } v \in V\}$.

Finally, we observe that $V^{\perp}$ is a weak* closed set in Y^* and that $(V^{\perp}, w^*)$ is homeomorphic with $((Y/V)^*, w^*)$. Moreover, the last space here belongs, by assumption, to S. Thus we have verified that Proposition 3.2.4 can be applied and therefore $Y \in \widetilde{S}$. $\square$

By putting together Theorems 3.2.2 and 3.2.3(i), (ii) we have that the class $\mathcal{A}$ of spaces isomorphic to a subspace of an Asplund generated space is included in $\widetilde{S}$. Thus we reproved Theorem 1.3.3.

We conclude by a useful equivalent characterization of the class $\widetilde{S}$. For its proof we need the following:

Proposition 3.2.5.

Let Z be a Baire space and $A \subset Z$ be a second-category set, that is, A is not first category. Then there exists a nonempty open set $U \subset Z$ such that the set $A \backslash U$ is first category and for every nonempty open set $G \subset U$ the set $A \cap G$ is second category. Moreover, $A \cap U$ is a Baire space.

Proof. Let $\mathcal{W}$ be the family of all open sets $W \subset Z$ such that $A \cap W$ is a first-category set. Put $\Omega = \bigcup \{W : W \in \mathcal{W}\}$ and $U = Z \backslash \overline{\Omega}$. We shall show that $A \cap \Omega$ is a first-category set. Then $A \backslash U$ will be a first-category set since

$$A \backslash U = A \cap \overline{\Omega} \subset (A \cap \Omega) \cup (\overline{\Omega} \backslash \Omega).$$

Thus the nonemptyness of U will also be guaranteed. That $A \cap G$ is a second-category set for every $\emptyset \neq G \subset U$ follows from the definition of the set U.

Let $\{W_{\gamma} : \gamma \in \Gamma\}$ be a disjoint subfamily of $\mathcal{W}$ which is maximal with respect to inclusion and put $\Omega_0 = \bigcup_{\gamma \in \Gamma} W_{\gamma}$. Then $\Omega \backslash \Omega_0$ is a nowhere dense set. In fact, assume that there is a nonempty open set $G \subset Z$ such that $G \subset \overline{\Omega \backslash \Omega_0}$. Then $G \subset Z \backslash \Omega_0$ and $G \cap \Omega \neq \emptyset$. Find $W \in \mathcal{W}$ such that $G \cap W \neq \emptyset$. Then the set $A \cap G \cap W$ is first category and so $G \cap W \in \mathcal{W}$. On the other hand, $G \cap W \cap \Omega_0 = \emptyset$, a contradiction with the maximality of the family $\{W_{\gamma} : \gamma \in \Gamma\}$.

Now we remark that

$$A \cap \Omega \subset (\Omega \backslash \Omega_0) \cup \bigcup_{\gamma \in \Gamma} (A \cap W_{\gamma});$$

thus it remains to show that the set $\bigcup_{\gamma \in \Gamma} (A \cap W_{\gamma})$ is first category. For every $\gamma \in \Gamma$ write $A \cap W_{\gamma} = \bigcup_{i=1}^{\infty} N_i^{\gamma}$, where the sets N_i^{γ} are nowhere dense. Since $N_i^{\gamma} \subset W_{\gamma}$ and $\{W_{\gamma} : \gamma \in \Gamma\}$ is a disjoint family of open sets, the sets $N_i := \bigcup_{\gamma \in \Gamma} N_i^{\gamma}$, $i = 1, 2, \ldots$, are easily seen to be nowhere dense. Therefore the set $\bigcup_{\gamma \in \Gamma} (A \cap W_{\gamma})$ $(= \bigcup_{i=1}^{\infty} N_i)$ is first category.

It remains to prove that $A \cap U$ is a Baire space. We first check that $\overline{A \cap U} \supset U$. Take an open set $W \subset U$ such that $W \cap (A \cap U) = \emptyset$; then $W \in \mathcal{W}$ and so $W \subset \Omega$. However, $W \subset U = Z \backslash \overline{\Omega}$, so $W \cap \Omega = \emptyset$, and therefore $W = \emptyset$.

Now, for $n = 1, 2, \ldots$, consider relatively open sets G_n in $A \cap U$ such that $\overline{G_n} \supset A \cap U$. We have to show that $\bigcap_{n=1}^{\infty} G_n \supset A \cap U$. For every $n \in$ IN we find an open set $H_n \subset U$ such that $G_n = A \cap H_n$. We note that

$$U \subset \overline{A \cap U} \subset \overline{G_n} \subset \overline{H_n}.$$

Hence, by putting $M = \bigcap_{n=1}^{\infty} H_n$, the set $U \setminus M$ is first category. It remains to prove that $\overline{A \cap M} \supset A \cap U$. So take any open set $\emptyset \neq W \subset U$. Assume for a while that $(A \cap M) \cap W = \emptyset$. Then $(A \cap W) \cap (M \cap W) = \emptyset$ and so $A \cap W \subset W \setminus M \ (\subset U \setminus M)$. Hence $A \cap W$ is a first-category set. However, from the first part of our proposition we know that the set $A \cap W$ is second category. Therefore $(A \cap M) \cap W \neq \emptyset$ and so $\overline{A \cap M} \supset U \ (\supset A \cap U)$, which means that $A \cap U$ is a Baire space.   □

Theorem 3.2.6.

A Banach space V belongs to the class $\widetilde{S}$ if (and only if) for each $\epsilon > 0$, each Baire space Z, and each minimal usco mapping $F : Z \to 2^{(V^*, w^*)}$ there exists a point $z \in Z$ such that the norm-diameter of Fz is less than ϵ.

Proof. Assume that a Banach space V does not belong to $\widetilde{S}$. Then there exist a Baire space Z and a minimal usco mapping $F : Z \to 2^{(V^*, w^*)}$ such that the set

$$Z_0 := \{z \in Z : Fz \text{ is not a singleton}\}$$

is second category. We can write $Z_0 = \bigcup_{n=1}^{\infty} Z_n$, where

$$Z_n := \left\{z \in Z : \text{norm-diameter of } Fz \text{ is greater than } \frac{1}{n}\right\}, \qquad n \in \text{IN}.$$

Let $n \in$ IN be such that Z_n is a second-category set; such an n surely exists. Let U be the open set found in Proposition 3.2.5 for $A = Z_n$. Thus $Z_n \cap U$ is a Baire space. Let Ψ denote the restriction of the mapping F to the space $Z_n \cap U$. Clearly, Ψ is usco. We shall verify, using Lemma 3.1.2, that Ψ is a minimal usco mapping. So let G be a relatively open set in $Z_n \cap U$ and let W be an open set in (V^*, w^*) such that $\Psi(G) \cap W \neq \emptyset$. Write $G = H \cap Z_n \cap U$, where H is an open set in Z. Then $F(H \cap U) \cap W \neq \emptyset$. Hence there exists a nonempty open set Ω in $H \cap U$ such that $F(\Omega) \subset W$. Now, by Proposition 3.2.5, we know that the set $Z_n \cap \Omega$ is second-category; hence in particular $Z_n \cap \Omega \neq \emptyset$. Thus $Z_n \cap \Omega$ is a nonempty relatively open set in $Z_n \cap U$ and $\Psi(Z_n \cap \Omega) \subset W$. This means that $\Psi : Z_n \cap U \to 2^{(V^*, w^*)}$ is a minimal usco mapping and this violates the condition of our theorem for $\epsilon = (1/n)$.   □

3.3. NOTES AND REMARKS

The explanation in Sections 3.1 and 3.2 follows papers of Stegall [St4, St5, St6] and that of Kenderov [K3]. The class $\mathcal{S}$ was originally defined via perfect mappings. For an account on perfect mappings see [En, Section 3.7].

While usco mappings have been around for a long time, see [Fr] and references therein, the concept of the minimal usco mapping was first considered, to our knowledge, by Christensen [Chr]. It should be noted that there exists a kind of duality between (minimal) usco mappings and (irreducible) perfect mappings.

Theorem 3.1.6 can be compared with Theorem 2.2.3: If a compact $K \in \mathcal{S}$, then we do not know if $C(K)$ is weak Asplund. Yet we get the same conclusion as in Theorem 2.2.3.

The second part of Theorem 3.2.2 can be generalized as follows: *If $g : Y \to V$ is a continuous mapping from a Banach space Y into V, $V \in \widetilde{\mathcal{S}}$, g is Gâteaux differentiable at the points of a residual set and $f : V \to \mathbb{R}$ is convex and continuous, then $f \circ g$ is Gâteaux differentiable at the points of a residual set.* This gives a new result even if $Y = V = \mathbb{R}$ [St4].

While the classes $\mathcal{S}$ and $\widetilde{\mathcal{S}}$ prove to be very useful in the theory of weak Asplund spaces, there are several open questions here. Let us mention at least one of them: *If a compact space K is in $\mathcal{S}$, does this imply that $(B_{C(K)^*}, w^*)$ is in $\mathcal{S}$?* See Theorems 1.2.4, 1.5.4, 5.2.7, and 7.1.10(v), which answer positively the same question for various subclasses of $\mathcal{S}$.

Another natural question, which has a positive answer, is whether the class $\mathcal{A}$ of Banach spaces isomorphic to a subspace of an Asplund generated space is a proper subclass of $\widetilde{\mathcal{S}}$. In fact, Talagrand's example [Ta2] of a weakly $\mathcal{K}$-analytic space from Section 4.3 lies in $\widetilde{\mathcal{S}} \backslash \mathcal{A}$ according to Theorems 4.1.2 and 4.3.1.

Theorem 3.2.3(vi) suggests the question of whether belonging to the class $\widetilde{\mathcal{S}}$ is a three-space property. That is, if $V \subset Y$, $V \in \widetilde{\mathcal{S}}$, and $Y/V \in \widetilde{\mathcal{S}}$, does it imply that $Y \in \widetilde{\mathcal{S}}$? The answer is negative, see Theorem 2.3.1.

Theorem 3.2.6 was suggested to us by W. B. Moors.

We conclude by a consequence of a "factorizing" result of Kenderov and Orihuela [KO]. *Let X be a compact space such that every minimal usco mapping from a complete metric space Y to X is single-valued at the points of a residual subset of Y. Then every minimal usco mapping from a regular Baire space Z to X is single-valued at the points of a residual subset of Z.*

Quite recently, Stegall proved, using a different approach, many other implications in this vein in the framework of completely regular spaces [St14]. Let us explain one of the ideas from this paper. Assume that a topological space X is such that for every Banach space V and for every minimal usco mapping $F : V \to 2^X$ there is a residual set in V where F is single-valued. Let K be a compact space and $\Phi : K \to 2^X$ be a minimal usco mapping. We shall show that Φ is single-valued on a residual set. Define $\partial : C(K) \to 2^K$ by $\partial(f) = \{k \in K : f(k) = \sup f(K)\}$. From Lemma 2.2.1 we know that this mapping is usco and open. From the proof of Theorem 3.1.6 we know that ∂ is a minimal usco mapping. Let us show that the composition $\Phi \circ \partial : C(K) \to 2^X$ is also a minimal usco mapping. We shall use Lemma 3.1.2 several times. So assume that $\Phi(\partial(\Omega)) \cap W \neq \emptyset$ for some open sets $\Omega \subset C(K)$ and $W \subset X$. Since the set $\partial(\Omega)$ is open and Φ is minimal usco, there is a nonempty open set $U \subset \partial(\Omega)$ such that $\Phi(U) \subset W$. Finally, as $\partial(\Omega) \cap U \neq \emptyset$ and ∂ is minimal usco, there is a nonempty open set $G \subset \Omega$ satisfying $\partial(G) \subset U$. Then $\Phi \circ \partial(G) \subset W$, which means that $\Phi \circ \partial$ is a minimal usco mapping. Next, the minimality of $\Phi \circ \partial$ and our assumption yield a residual set $\Omega \subset C(K)$ such that $\Phi(\partial(f))$ is a singleton for every $f \in \Omega$. Then $\Phi(k)$ will be a singleton for every $k \in \partial(\Omega)$. Finally, playing a simple Banach–Mazur game, we can prove that $\partial(\Omega)$ is a residual set; see Lemma 4.2.1.

Chapter Four

Two More Concrete Classes of Banach Spaces That Lie in $\widetilde{S}$

Chapter 3 provided us with quite a rich theory for Stegall's class $\widetilde{S}$. However, at the end of this chapter, the question of which Banach spaces, besides subspaces of Asplund generated spaces, belong to $\widetilde{S}$ remained unanswered. In the present chapter, we show that there are two further concrete classes of Banach spaces which belong to $\widetilde{S}$: weakly $\mathcal{K}$-analytic spaces and Gâteaux smooth spaces. Then a counterexample of a weakly $\mathcal{K}$-analytic space which is not a subspace of a WCG space is presented.

4.1. WEAKLY $\mathcal{K}$-ANALYTIC SPACES BELONG TO $\widetilde{S}$

The space $\mathbb{IN}^{\mathbb{IN}}$ is considered with the product topology. Thus there exists a metric ρ which generates the topology on $\mathbb{IN}^{\mathbb{IN}}$ and $(\mathbb{IN}^{\mathbb{IN}}, \rho)$ is a complete separable metric space. Let us denote

$$D = \{\emptyset\} \cup \{0, 1\} \cup \{0, 1\}^2 \cup \ldots,$$

that is, D is the set of finite sequences of zeros and ones. For $d \in D$ let $|d|$ denote the cardinality of d. In particular, $|\emptyset| = 0$. If $d, d' \in D$, the relation $d \succ d'$ says that $|d| > |d'|$ and d' consists of the first $|d'|$ elements of d. If $d = \{i_1, i_2, \ldots, i_n\} \in D$ and $i \in \{0, 1\}$, we put $di = \{i_1, i_2, \ldots, i_n, i\}$.

Definition 4.1.1. A completely regular space is called $\mathcal{K}$-*analytic* if there exists a usco mapping from $\mathbb{IN}^{\mathbb{IN}}$ onto it. A Banach space V is called *weakly $\mathcal{K}$-analytic* if (V, w) is $\mathcal{K}$-analytic.

Here it should be noted that any Banach space V endowed with the weak topology is completely regular; see the argument below Definition 3.2.1.

Theorem 4.1.2.

Every weakly $\mathcal{K}$-analytic space belongs to class $\widetilde{S}$.

Proof. Let V be a weakly $\mathcal{K}$-analytic space. Then we can easily construct a usco mapping φ from $\mathrm{IN}^{\mathrm{IN}}$ onto the unit ball (B_V, w). We shall assume that V does not belong to $\widetilde{S}$. Then, by Theorem 3.2.6, there exist $\epsilon > 0$, a Baire space Z and a minimal usco mapping $F : Z \to 2^{(V^*, w^*)}$ such that the norm diameter of Fz is greater than ϵ for every $z \in Z$.

For every $d \in D$ we shall construct a second-category set Z_d in Z, an open set Σ_d in $(\mathrm{IN}^{\mathrm{IN}}, \rho)$, of diameter not greater than $\frac{1}{|d|}$, and an element $v_d \in \varphi(\Sigma_d)$ such that

$$Z_d \subset Z_{d'} \quad \text{and} \quad \Sigma_d \subset \Sigma_{d'} \quad \text{if} \quad d \succ d',$$

$$\mathrm{diam}_{\varphi(\Sigma_d)} Fz := \sup \left\{ \langle \xi_1 - \xi_0, v \rangle : \ \xi_0, \xi_1 \in Fz, \ v \in \varphi(\Sigma_d) \right\} > \epsilon$$

for every $z \in Z_d$, and

$$\inf \langle F(Z_{d1}) - F(Z_{d0}), v_d \rangle > \epsilon. \qquad (*)$$

Consider the sets $Z_m = \{ z \in Z : \ Fz \cap mB_{V^*} \neq \emptyset \}$, $m \in \mathrm{IN}$. Each Z_m is closed since F is upper semicontinuous and mB_{V^*} is weak* closed. As Z is a Baire space, there is $m \in \mathrm{IN}$ such that the set Z_m has nonempty interior. Denote this interior by $Z_\emptyset$. Clearly, $Z_\emptyset$ is a second category set. Moreover, Lemma 3.1.2(ii) guarantees that $F(Z_\emptyset) \subset mB_{V^*}$. Further put $\Sigma_\emptyset = \mathrm{IN}^{\mathrm{IN}}$.

Assume that we have already constructed Z_d, Σ_d for some $d \in D$. Let U be the set found in Proposition 3.2.5 for $A = Z_d$. Take some $z \in Z_d \cap U$. We find $v_d \in \varphi(\Sigma_d)$ and $\xi_0, \xi_1 \in Fz$ such that $\langle \xi_1 - \xi_0, v_d \rangle > \epsilon$. Let $a > 0$ be so small that $\langle \xi_1 - \xi_0, v_d \rangle > \epsilon + a$. Consider the set $W = \{ \xi \in V^* : \ \langle \xi, v_d \rangle > \langle \xi_0, v_d \rangle + \epsilon + a \}$. This set is weak* open and $F(U) \cap W \ni \xi_1$. Hence, by Lemma 3.1.2(iii), there is a nonempty open set $U_1 \subset U$ such that $F(U_1) \subset W$ and so

$$\inf \langle F(U_1), v_d \rangle \geq \langle \xi_0, v_d \rangle + \epsilon + a.$$

By a similar argument we obtain a nonempty open set $U_0 \subset U$ such that

$$\inf \langle F(U_1), v_d \rangle > \sup \langle F(U_0), v_d \rangle + \epsilon.$$

Since Σ_d is an open set in the separable metric space, we can write $\Sigma_d = \bigcup_{n=1}^{\infty} \Sigma^n$, where each Σ^n is an open set of diameter not greater than $(|d| + 1)^{-1}$. Then $Z_d \cap U_1 = \bigcup_{n=1}^{\infty} Z^n$, where

$$Z^n = \{ z \in Z_d \cap U_1 : \ \mathrm{diam}_{\varphi(\Sigma^n)} Fz > \epsilon \}, \qquad n \in \mathrm{IN}.$$

Let $n \in \mathrm{IN}$ be such that the set Z^n is second category. Such an n exists since the set $Z_d \cap U_1$ is second category by Proposition 3.2.5. Then put $Z_{d1} = Z^n$ and $\Sigma_{d1} = \Sigma^n$. The sets Z_{d0} and Σ_{d0} are constructed in an analogous way. This finishes the induction step. We can immediately check that Z_d, Σ_d, and v_d constructed for every $d \in D$ in the above way have all the proclaimed properties.

Denote $K_0 = \{v_d : d \in D\}$. We shall show that this set is relatively weakly compact. To this end, for every $d \in D$ we find $\sigma_d \in V_d$ such that $v_d \in \varphi(\sigma_d)$. Denote $C_0 = \{\sigma_d : d \in D\}$; thus $K_0 \subset \varphi(C_0)$. We shall first show that C_0 is a relatively compact set in (the complete metric space) $(\mathbb{N}^{\mathbb{N}}, \rho)$. So fix any $\Delta > 0$ and put $D_0 = \{d \in D : |d| \leq 1/\Delta + 1\}$; this is a finite set. Then $\{\sigma_d : d \in D_0\}$ is a (finite) Δ-net for the set C_0. Indeed, if $d \in D$ and $|d| > 1/\Delta + 1$, then there exists $d_0 \in D_0$ such that $d \succ d_0$ and $1/\Delta < |d_0|$. It then follows that $\sigma_d \in \Sigma_d \subset \Sigma_{d_0}$ and so the distance between σ_d and σ_{d_0} is not greater than $|d_0|^{-1} < \Delta$. Now [En, Theorem 4.3.29] guarantees that C_0 is a relatively compact set.

Let C denote the closure of C_0; thus C is a compact set in $\mathbb{N}^{\mathbb{N}}$. Then $K_0 \subset \varphi(C)$ and it remains to show that the set $\varphi(C)$ is weakly compact. In this way, the relative weak compactness of K_0 will be proved. So suppose $\varphi(C) \subset \bigcup_{\gamma \in \Gamma} U_\gamma$, where each U_γ is a weakly open subset of V. Since φ is compact valued, for each $\sigma \in C$ there is a finite subset $\Gamma_\sigma \subset \Gamma$ such that $\varphi(\sigma) \subset \bigcup_{\gamma \in \Gamma_\sigma} U_\gamma =: W_\sigma$. Since φ is upper semicontinuous, for each $\sigma \in C$ there is an open set $\sigma \in \Omega_\sigma \subset \mathbb{N}^{\mathbb{N}}$ such that $\varphi(\Omega_\sigma) \subset W_\sigma$. Then $C \subset \bigcup_{\sigma \in C} \Omega_\sigma$. Now, C is compact; therefore there exists a finite set $E \subset C$ satisfying $C \subset \bigcup_{\sigma \in E} \Omega_\sigma$. Thus,

$$\varphi(C) \subset \bigcup_{\sigma \in E} \varphi(\Omega_\sigma) \subset \bigcup_{\sigma \in E} W_\sigma = \bigcup_{\sigma \in E} \bigcup_{\gamma \in \Gamma_\sigma} U_\gamma.$$

Hence, $\varphi(C)$ is a weakly compact set.

Next, for each $\delta = \{i_1, i_2, \ldots\} \in \{0, 1\}^{\mathbb{N}}$ we find ξ_δ in $\bigcap_{n=1}^{\infty} \overline{F(Z_{\{i_1, i_2, \ldots, i_n\}})}^*$; this is possible as $F(Z) \subset m B_{V^*}$. (Here $\overline{M}^*$ means the weak* closure of $M \subset V^*$.) Then for any $\gamma, \delta \in \{0, 1\}^{\mathbb{N}}$, with $\gamma \neq \delta$, we have by $(*)$ that

$$\rho_{K_0}(\xi_\gamma, \xi_\delta) := \sup \{|\langle \xi_\gamma - \xi_\delta, v \rangle| : v \in K_0\} \geq \epsilon.$$

Thus, by observing that the set $\{\eta_\delta : \delta \in \{0, 1\}^{\mathbb{N}}\}$ is uncountable, we conclude that the (pseudometric) space (V^*, ρ_{K_0}) is not separable. This means, by Definition 1.4.1, that K_0 is not an Asplund set. However, $K_0 \subset \varphi(C)$ and $\varphi(C)$ is weakly compact. Hence, by Proposition 1.4.2, K_0 is an Asplund set. This gives us our desired contradiction. □

By putting together Theorem 3.2.2 and the just proved theorem, we get:

Corollary 4.1.3.

Weakly $\mathcal{K}$-analytic spaces are weak Asplund.

4.2. *GÂTEAUX SMOOTH SPACES BLONG TO $\tilde{S}$*

Let X be a topological space. In order to prove that some subset of X is residual, that is, that its complement is a first-category set, we may use the so-called *Banach–Mazur game*. What does this mean? Let X be a topological space and M its subset which is suspected of being residual. We consider two players **A** and **B**. A Banach–Mazur play runs as follows. Both players choose, alternatively, nonempty open subsets of X. The play is always started by player **A**, who moves $U_1 \subset X$. Then player **B** moves $W_1 \subset U_1$. Further $U_2 \subset W_1$ is player **A**'s answer to W_1, and $W_2 \subset U_2$ is the move of player **B**. The play continues in an obvious manner. We say that player **B** has *won* the play if $\bigcap_{k=1}^{\infty} W_k$ is a (maybe empty) subset of the set M. Further, we say that player **B** *has a winning strategy* if he is able to win every play, independently of player **A**'s moves.

Lemma 4.2.1.

*The set $M \subset X$ is residual if and only if player **B** has a winning strategy.*

Proof. Assume that there are dense open sets $G_n \subset X$, $n = 1, 2, \ldots$, such that $\bigcap_{n=1}^{\infty} G_n \subset M$. Let $\Omega \subset X$ be any open set. If $\Omega \subset \bigcap_{n=1}^{\infty} G_n$, we put $f(\Omega) = \Omega$. Otherwise put $f(\Omega) = \Omega \cap G_m$, where m is the first $n \in \text{IN}$ such that Ω is not a subset of G_n. Let us now play the Banach–Mazur play and assume that the k-th move of player **A** was U_k. Then a winning strategy for player **B** is $W_k = f(U_k)$. In fact, then surely $\bigcap_{k=1}^{\infty} W_k \subset \bigcap_{k=1}^{\infty} G_k \subset M$.

Conversely, let player **B** have a winning strategy $f_1, f_2, \ldots$. That is, f_1 is a recipe of how to choose W_1 if U_1 is known (i.e., $W_1 = f_1(U_1)$), and for a general $k \in \text{IN}$ we have $W_k = f_k(U_1, W_1, \ldots, U_{k-1}, W_{k-1}, U_k)$. In what follows, we shall also put $f_k(U_1, W_1, \ldots, U_{k-1}, W_{k-1}, \emptyset) = \emptyset$. Let ξ be an ordinal whose cardinality is greater than the cardinality of the set X. Put $U_1 = X$ and $W_1 = f_1(U_1)$. Let $\gamma \in (1, \xi)$ be a fixed ordinal and assume we have found U_α, W_α for all $\alpha < \gamma$. Put $U_\gamma = \text{int}(X \setminus \bigcup_{\alpha < \gamma} W_\alpha)$ and $W_\gamma = f_1(U_\gamma)$. Clearly, for some $\gamma < \xi$ we get that $U_\gamma = W_\gamma = \emptyset$; so $\bigcup_{\alpha < \xi} W_\alpha$ is dense in X. Also, we can see that $W_\alpha \cap W_\beta = \emptyset$ if $\alpha \neq \beta$. Now fix any $\alpha < \xi$ for which $W_\alpha \neq \emptyset$. Put $U_{\alpha 1} = W_\alpha$ and $W_{\alpha 1} = f_2(U_\alpha, W_\alpha, U_{\alpha 1})$. Assume we know $U_{\alpha\beta}, W_{\alpha\beta}$ for all β less than some fixed $\gamma \leq \xi$. Then put $U_{\alpha\gamma} = \text{int}(W_\alpha \setminus \bigcup_{\beta < \gamma} W_{\alpha\beta})$ and $W_{\alpha\gamma} = f_2(U_\alpha, W_\alpha, U_{\alpha\gamma})$. Clearly, we eventually get that $\bigcup_{\beta < \xi} W_{\alpha\beta}$ is dense in W_α and so $\bigcup \{ W_{\alpha\beta} : \alpha, \beta < \xi \}$ is dense in X. Observe also that $W_{\alpha\beta} \cap W_{\gamma\delta} = \emptyset$ if $\alpha \neq \gamma$ or $\beta \neq \delta$. Continuing this process, we obtain open sets $U_{\alpha_1 \ldots \alpha_k}$, $W_{\alpha_1 \ldots \alpha_k}$ in X, $\alpha_1, \alpha_2, \ldots \leq \xi$, $k = 1, 2, \ldots$, such that

$$W_{\alpha_1 \ldots \alpha_k} = f_k \left(U_{\alpha_1}, \ W_{\alpha_1}, \ U_{\alpha_1 \alpha_2}, \ W_{\alpha_1 \alpha_2}, \ \ldots, \ U_{\alpha_1 \ldots \alpha_{k-1}}, \ W_{\alpha_1 \ldots \alpha_{k-1}}, \ U_{\alpha_1 \ldots \alpha_k} \right)$$

and that $\Omega_k := \bigcup \{W_{\alpha_1 \ldots \alpha_k} : \alpha_1, \ldots, \alpha_k < \xi\}$ is dense in X for each $k \in \mathbb{N}$. Moreover, $W_{\alpha_1 \ldots \alpha_{k-1}} \supset W_{\alpha_1 \ldots \alpha_k}$ and for each fixed $k \in \mathbb{N}$ the sets $W_{\alpha_1 \ldots \alpha_k}$ are mutually disjoint. Now take any x in $\bigcap_{k=1}^{\infty} \Omega_k$ (if any). Then there exists a sequence $\{\alpha_k\}$ such that $x \in W_{\alpha_1} \cap W_{\alpha_1 \alpha_2} \cap \cdots$. We realize that the sequence $U_{\alpha_1}, W_{\alpha_1}, U_{\alpha_1 \alpha_2}, W_{\alpha_1 \alpha_2}, \ldots$ can be viewed as a Banach–Mazur play where $W_{\alpha_1}, W_{\alpha_1 \alpha_2}, \ldots$ have been chosen according to the winning strategy $f_1, f_2, \ldots$ of player **B**. Therefore x belongs to M. We have thus proved that $\bigcap_{k=1}^{\infty} \Omega_k \subset M$. As each Ω_k is dense in X, the set M is residual. $\qquad\square$

Before proving the main result of this section we shall consider a simpler situation.

Proposition 4.2.2.

Let V be a Banach space and F be a minimal usco mapping from a Baire space Z into (V^*, w^*). Then the set M of all $z \in Z$ such that the convex hull $\mathrm{co} Fz$ of Fz lies on a sphere around the origin in V^* is residual.

Proof. We shall consider a Banach–Mazur play to show that the set M is residual. Let $\{\epsilon_k\} \subset (0,1)$ be a sequence going to 0. Let player **A** choose $U_1 \subset Z$. Consider sets $C_n = \{z \in U_1 : Fz \cap nB_{V^*} \neq \emptyset\}$, $n = 1, 2, \ldots$. These sets are closed in U_1 since F is upper semicontinuous, and their union is U_1. It then follows from the definition of Baire space that there are $n \in \mathbb{N}$ and a nonempty open set $W_1 \subset C_n$. This W_1 will be player **B**'s answer to U_1. Thus, by Lemma 3.1.2(ii), $F(W_1)$ is a subset of nB_{V^*}.

Now let U_2 be player **A**'s answer to W_1. Put $s_2 = \sup \{\|\xi\| : \xi \in F(U_2)\}$. Consider first $s_2 = 0$. Then it is easy to see how player **B** will win the play. Indeed, it suffices to put $W_k = U_k$ for each $k \geq 2$. Next, consider $s_2 > 0$. We find $\xi \in F(U_2)$ such that $\|\xi\| > s_2(1 - \epsilon_2)$. Also, pick $e_2 \in V$, $\|e_2\| = 1$, satisfying $\langle \xi, e_2 \rangle > s_2(1 - \epsilon_2)$. Thus, denoting $\Omega = \{v^* \in V^* : \langle v^*, e_2 \rangle > s_2(1 - \epsilon_2)\}$, we have $F(U_2) \cap \Omega \neq \emptyset$. Now, since Ω is weak* open, Lemma 3.1.2(iii) gives us a nonempty open set $W_2 \subset U_2$ such that $F(W_2) \subset \Omega$; thus

$$\inf \langle F(W_2), e_2 \rangle := \inf \{\langle v^*, e_2 \rangle : v^* \in F(W_2)\} \geq s_2(1 - \epsilon_2) > 0.$$

This W_2 will be the second move of player **B**.

Repeating this process in all steps, we obtain nonempty open sets $U_1, U_2, \ldots, W_1, W_2, \ldots$ in X, positive numbers $s_1, s_2, \ldots$, and norm 1 elements $e_1, e_2, \ldots$ in V such that

$$U_1 \supset W_1 \supset U_2 \supset W_2 \supset \ldots, \qquad s_k = \sup \{\|\xi\| : \xi \in F(U_k)\},$$

and

$$\inf \langle F(W_k), e_k \rangle \geq s_k(1 - \epsilon_k) > 0$$

for $k = 2, 3, \ldots$. If $\bigcap_{k=1}^{\infty} W_k$ is empty, then player **B** wins the play and we are done. Otherwise, take any z in this intersection. Remarking that $s_2 \geq s_3 \geq \cdots > 0$, we put $s = \lim_k s_k$. We shall show that $\|\xi\| = s$ for every ξ in coFz, which will finish the proof. So let $\xi_1, \ldots, \xi_n \in Fz$ and $a_1, \ldots, a_n > 0$, with $a_1 + \cdots + a_n = 1$, be given. Since $\xi_i \in Fz \subset F(W_k) \subset F(U_k)$, we have

$$s_k \geq a_1 \|\xi_1\| + \cdots + a_n \|\xi_n\| \geq \|a_1 \xi_1 + \cdots + a_n \xi_n\|$$
$$\geq \langle a_1 \xi_1 + \cdots + a_n \xi_n, e_k \rangle \geq s_k (1 - \epsilon_k)$$

for all $k \in \text{IN}$. By letting k go to infinity, we get $s = \|a_1 \xi_1 + \cdots + a_n \xi_n\|$. Therefore, $\bigcap_{k=1}^{\infty} W_k \subset M$, and so the set M is residual. $\qquad\square$

We recall that a norm is strictly convex if and only if the only convex objects in spheres are singletons. Thus, we immediately obtain the next theorem.

Theorem 4.2.3.

A Banach space belongs to Stegall's class $\tilde{S}$ if its dual admits an equivalent strictly convex dual norm.

The aim of this section is to show that the assumption in the above theorem can be slightly weakened.

Theorem 4.2.4.

A Banach space belongs to class $\tilde{S}$ if it admits an equivalent Gâteaux smooth norm.

Proof. Let Z be a Baire space, let V be a Banach space with a Gâteaux smooth norm $\|\cdot\|$ and let $F : Z \to 2^{(V^*, w^*)}$ be a minimal usco mapping. We denote by M the set of all $z \in Z$ for which Fz is a singleton. Our task is to show that M is residual. We shall do this by playing a Banach–Mazur play associated with the set M. A rough winning strategy of player **B** will be as that in the proof of Proposition 4.2.2. We realize a crucial fact: If $z \in Z$ and $s > 0$ are such that Fz lies on a sphere of radius s and center 0 and, moreover, there exists $e \in V$, $\|e\| = 1$, such that $\langle \xi, e \rangle = s$ for all ξ from Fz, then Fz must be a singleton (owing to the smoothness of $\|\cdot\|$). Since we are afraid that such an e will not always exist, player **B** must proceed more carefully in order to win any play. In fact, we shall have to change the original norm, step by step, so that, at infinity, we shall obtain an equivalent Gâteaux smooth norm $\|\cdot\|_\infty$ and $e \in V$, $\|e\|_\infty = 1$, satisfying $\langle \xi, e \rangle = \|\xi\|_\infty = $ const. for each $\xi \in Fz \in Z$. Thus, Fz will be shown to be a singleton.

Choose sequences $\{\epsilon_k\}, \{\beta_k\}$ of positive numbers such that

$$\tfrac{1}{2} > \epsilon_2 > \epsilon_3 > \cdots, \quad \sum_{k=2}^{\infty} \beta_k^2 < 3, \quad \text{and} \quad \sum_{k=2}^{\infty} \frac{\sqrt{\epsilon_k}}{\beta_k} < +\infty;$$

then $\epsilon_k \to 0$ as $k \to \infty$. Let W_1 be the same answer of player **B** to U_1 as it was in the proof of Proposition 4.2.2. Further, let player **A** move $U_2 \subset W_1$. Put $s_2 = \sup\{\|\xi\|_2 : \xi \in F(U_2)\}$, where $\|\cdot\|_2$ means the original norm $\|\cdot\|$. Here and below we shall use the same symbol for a norm on V and the corresponding dual norm on V^*. If $s_2 = 0$, we already know how **B** should move in order to win the play. So further assume $s_2 > 0$. As in the proof of Proposition 4.2.2, we find $e_2 \in V$, $\|e_2\|_2 = 1$, and player **B** finds a nonempty open set $W_2 \subset U_2$ such that $\inf\langle F(W_2), e_2\rangle \geq s_2(1 - \epsilon_2)$. Now suppose player **A** chooses $U_3 \subset W_2$. We define a norm $\|\cdot\|_3$ on V by $\|\cdot\|_3^2 = \|\cdot\|_2^2 + \beta_2^2 \operatorname{dist}(\cdot, \mathbb{R}e_2)^2$. Here $\operatorname{dist}(\cdot, \mathbb{R}e_2)$ means $\inf\{\|\cdot - \lambda e_2\| : \lambda \in \mathbb{R}\}$. It is easy to verify that $\|\cdot\|_3$ is an equivalent Gâteaux smooth norm on V and that $\|e_2\|_3 = 1$. Put $s_3 = \sup\{\|\xi\|_3 : \xi \in F(U_3)\}$ and find $e_3 \in V$, $\|e_3\|_3 = 1$, and a nonempty open set $W_3 \subset U_3$ satisfying $\inf\langle F(W_3), e_3\rangle \geq s_3(1 - \epsilon_3)$. This is possible by Lemma 3.1.2(iii). We remark that $s_3 > 0$. Indeed, since $\|e_2\|_3 = 1$, and $U_3 \subset W_2$, we have

$$s_3 \geq \sup\langle F(U_3), e_2\rangle \geq \inf\langle F(U_3), e_2\rangle$$
$$\geq \inf\langle F(W_2), e_2\rangle \geq s_2(1 - \epsilon_2) > 0.$$

In the subsequent steps player **B** will use a similar strategy. This means, once U_k, W_k, $\|\cdot\|_k$, s_k, and e_k are known and player **A**'s answer to W_k is U_{k+1}, we define

$$\|\cdot\|_{k+1}^2 = \|\cdot\|_k^2 + \beta_k^2 \operatorname{dist}(\cdot, \mathbb{R}e_k)^2 \quad \text{on} \quad V,$$

$$s_{k+1} = \sup\{\|\xi\|_{k+1} : \xi \in F(U_{k+1})\}$$

and player **B** finds $e_{k+1} \in V$, $\|e_{k+1}\|_{k+1} = 1$, and a nonempty open set $W_{k+1} \subset U_{k+1}$ such that

$$\inf\langle F(W_{k+1}), e_{k+1}\rangle \geq s_{k+1}(1 - \epsilon_{k+1}).$$

Assuming that the norm $\|\cdot\|_k$ is equivalent and Gâteaux smooth, then so is $\|\cdot\|_{k+1}$. Further, we know that $s_k > 0$, which implies that s_{k+1} is also positive.

Assume the above Banach–Mazur play has been finished. If $\bigcap_{k=1}^{\infty} W_k$ is empty, then we are done. So suppose that this intersection is nonempty. We observe that on V

$$\|\cdot\|^2 \leq \|\cdot\|_k^2 \leq \|\cdot\|_{k+1}^2 \leq \left(1 + \sum_{j=2}^{k} \beta_j^2\right)\|\cdot\|^2$$

for all $k \geq 2$. Thus, $\| \cdot \|_k$ converges, uniformly on bounded sets, to some $\| \cdot \|_\infty$ and so $\| \cdot \|_\infty$ will be an equivalent norm on V. More precisely, we have $\| \cdot \| \leq \| \cdot \|_\infty \leq 2\| \cdot \|$. We shall show that $\| \cdot \|_\infty$ *is Gâteaux smooth.* So take $0 \neq v \in V$, $h \in V$, and $t > 0$. Since $\mathrm{dist}(\cdot, \mathrm{IR}e_k)$ is a seminorm, we have, for all $k \in \mathrm{IN}$,

$$\mathrm{dist}(v + th, \mathrm{IR}e_k)^2 + \mathrm{dist}(v - th, \mathrm{IR}e_k)^2 - 2\,\mathrm{dist}(v, \mathrm{IR}e_k)^2$$
$$\leq 4t\,\mathrm{dist}(h, \mathrm{IR}e_k)\mathrm{dist}(v, \mathrm{IR}e_k) + 2t^2\mathrm{dist}(h, \mathrm{IR}e_k)^2$$
$$\leq 4t\|h\|\|v\| + 2t^2\|h\|^2$$

and thus

$$\frac{1}{t}\left(\|v + th\|_\infty^2 + \|v - th\|_\infty^2 - 2\|v\|_\infty^2 \right)$$
$$= \frac{1}{t}\left(\|v + th\|_n^2 + \|v - th\|_n^2 - 2\|v\|_n^2 \right)$$
$$+ \frac{1}{t}\sum_{k=n}^{\infty} \beta_k^2 \left(\mathrm{dist}(v + th, \mathrm{IR}e_k)^2 + \mathrm{dist}(v - th, \mathrm{IR}e_k)^2 - 2\,\mathrm{dist}(v, \mathrm{IR}e_k)^2 \right)$$
$$\leq \frac{1}{t}\left(\|v + th\|_n^2 + \|v - th\|_n^2 - 2\|v\|_n^2 \right) + \sum_{k=n}^{\infty} \beta_k^2 \left(4\|h\|\|v\| + 2t\|h\|^2 \right).$$

Here $\| \cdot \|_n$ is Gâteaux differentiable at v, so

$$\lim_{t\downarrow 0} \frac{1}{t}\left(\|v + th\|_\infty^2 + \|v - th\|_\infty^2 - 2\|v\|_\infty^2 \right) \leq 4\|h\|\|v\| \sum_{k=n}^{\infty} \beta_k^2.$$

However, the right-hand side here goes to 0 as $n \to \infty$. This shows that $\| \cdot \|_\infty^2$ and hence $\| \cdot \|_\infty$ is Gâteaux differentiable at v. Let us also remark that $0 < s_{k+1} \leq s_k$ because $U_{k+1} \subset U_k$ and $\|\xi\|_{k+1} \leq \|\xi\|_k$ for all $\xi \in V^*$; thus s_k converge to some s_∞.

Further, we claim that *the sequence $\{e_k\}$ converges in norm to some $e_\infty \in V$.* Assume, for a moment, the claim has been verified. Then

$$\left| \|e_\infty\|_\infty - 1 \right| \leq \left| \|e_\infty\|_\infty - \|e_k\|_\infty \right| + \left| \|e_k\|_\infty - \|e_k\|_k \right|$$
$$\leq \|e_\infty - e_k\|_\infty + \|e_k\|_\infty - \|e_k\|_k \to 0 \quad \text{as} \quad k \to +\infty,$$

so that $\|e_\infty\|_\infty = 1$. Now take any z in $\bigcap_{k=1}^{\infty} W_k$ (if any) and any ξ in Fz. Then for all $k \geq 2$ we have $\xi \in F(W_k) \subset F(U_k)$ and so

$$s_k(1 - \epsilon_k) \leq \inf\langle F(W_k), e_k \rangle \leq \langle \xi, e_k \rangle \leq \|\xi\|_k \leq s_k.$$

Therefore $s_\infty \leq \langle \xi, e_\infty \rangle \leq \|\xi\|_\infty \leq s_\infty$ for all $\xi \in Fz$. Now the Gâteaux smoothness of $\| \cdot \|_\infty$ enters and therefore Fz must be a singleton; thus $z \in M$. (We have in fact proved that $F\left(\bigcap_{k=1}^{\infty} W_k \right)$ must be a singleton.)

It remains to prove the claim. Fix any $k \geq 2$. Choose $\xi \in F(W_{k+1})$. Then, recalling that $W_{k+1} \subset U_{k+1} \subset W_k \subset U_k$ and $\|e_k\|_{k+1} = 1$, we have

$$s_k \|e_{k+1}\|_k \geq \|\xi\|_k \|e_{k+1}\|_k \geq \langle \xi, e_{k+1} \rangle \geq s_{k+1}(1 - \epsilon_{k+1})$$
$$\geq \|\xi\|_{k+1}(1 - \epsilon_{k+1}) \geq \langle \xi, e_k \rangle (1 - \epsilon_{k+1})$$
$$\geq s_k(1 - \epsilon_k)(1 - \epsilon_{k+1}) > s_k(1 - 2\epsilon_k);$$
$$\|e_{k+1}\|_k > (1 - 2\epsilon_k).$$

Now find $\lambda_k \in \mathbb{R}$ such that $\mathrm{dist}(e_{k+1}, \mathbb{R}e_k) = \|e_{k+1} - \lambda_k e_k\|$. The last inequality implies

$$1 = \|e_{k+1}\|_{k+1}^2 = \|e_{k+1}\|_k^2 + \beta_k^2 \|e_{k+1} - \lambda_k e_k\|^2$$
$$> (1 - 2\epsilon_k)^2 + \beta_k^2 \|e_{k+1} - \lambda_k e_k\|^2;$$
$$\|e_{k+1} - \lambda_k e_k\| < 2\sqrt{\epsilon_k}/\beta_k.$$

Thus

$$\big|1 - |\lambda_k|\big| = \big|\|e_{k+1}\|_{k+1} - \|\lambda_k e_k\|_{k+1}\big| \leq \|e_{k+1} - \lambda_k e_k\|_{k+1} < 4\sqrt{\epsilon_k}/\beta_k.$$

Finally, for $\xi \in F(W_{k+1})$ $(\subset F(U_{k+1}) \subset F(W_k))$ we have

$$\lambda_k \langle \xi, e_k \rangle = \langle \xi, e_{k+1} \rangle - \langle \xi, e_{k+1} - \lambda_k e_k \rangle$$
$$\geq s_{k+1}(1 - \epsilon_{k+1}) - \|\xi\|_{k+1} \|e_{k+1} - \lambda_k e_k\|_{k+1}$$
$$> s_{k+1}(1 - \epsilon_{k+1}) - s_{k+1} 4\sqrt{\epsilon_k}/\beta_k$$

and $\langle \xi, e_k \rangle \geq s_k(1 - \epsilon_k) > 0$. It follows that for large $k \in \mathbb{N}$ we have $\lambda_k > 0$ and so $|1 - \lambda_k| < 4\sqrt{\epsilon_k}/\beta_k$. Now we can conclude

$$\|e_{k+1} - e_k\| \leq \|e_{k+1} - \lambda_k e_k\| + \|\lambda_k e_k - e_k\| < 2\sqrt{\epsilon_k}/\beta_k + |\lambda_k - 1| < 6\sqrt{\epsilon_k}/\beta_k.$$

As $\sum_{k=1}^{\infty} \sqrt{\epsilon_k}/\beta_k$ is finite, we have shown that $\{e_k\}$ is a Cauchy sequence. $\square$

Corollary 4.2.5.

Banach spaces that admit an equivalent Gâteaux smooth norm are weak Asplund.

Proof. Put together Theorems 4.2.4 and 3.2.2.

4.3. A WEAKLY $\mathcal{K}$-ANALYTIC SPACE MAY NOT BE A SUBSPACE OF A WCG SPACE

Put $\Gamma = \mathbb{N}^{\mathbb{N}}$. For $n = 1, 2, \ldots$ we let $\mathcal{A}_n$ be the family from Section 1.6, that is, the family of all subsets $A \subset \Gamma$ with the following property: If $\sigma, \rho \in A$ and $\sigma \neq \rho$, then $\sigma|n = \rho|n$ and $\sigma(n+1) \neq \rho(n+1)$. Thus every $A \in \mathcal{A}_n$ is at most countable. Denote $\mathcal{A} = \bigcup_{n=1}^{\infty} \mathcal{A}_n$ and define $K = \{\chi_A : A \in \mathcal{A}\}$, where χ_A means the characteristic function of the set A. It is a routine matter to check that K is a closed subspace in $\{0,1\}^{\Gamma}$ and hence compact. In fact, consider a net $\{\chi_{A_\tau}\}$ with $A_\tau \in \mathcal{A}$ converging to $f \in \{0,1\}^{\Gamma}$. Clearly, $f = \chi_A$ for some set $A \subset \Gamma$. If A is a singleton, then we are done. Otherwise take any two distinct elements σ, ρ in A. Then there surely exists $n \in \mathbb{N}$ such that $\sigma|n = \rho|n$ and $\sigma|(n+1) \neq \rho|(n+1)$. We have to show that $A \in \mathcal{A}_n$. So take any two distinct elements $\sigma_1, \sigma_2 \in A$. We find τ so that $\sigma_1, \sigma_2, \sigma, \rho \in A_\tau$. Then, necessarily, $A_\tau \in \mathcal{A}_n$ and so $\sigma_1|n = \sigma_2|n$ and $\sigma_1|(n+1) \neq \sigma_2|(n+1)$. Hence, $A \in \mathcal{A}_n \subset \mathcal{A}$ and $f = \chi_A \in K$.

Theorem 4.3.1.

The space $C(K)$, with K defined above is a weakly $\mathcal{K}$-analytic space yet K is not an Eberlein compact, and so $C(K)$ is not isomorphic to a subspace of any WCG space or even to a subspace of any Asplund generated space.

     Proof. For $\sigma \in \Gamma$ we define $\pi_\sigma : K \to \mathbb{R}$ by $\pi_\sigma(k) = k(\sigma)$, $k \in K$. Clearly, $\pi_\sigma \in C(K)$. Now we define $\varphi : \Gamma \to 2^{C(K)}$ as

$$\varphi(\sigma) = \{\pi_\sigma, 0\}, \qquad \sigma \in \Gamma.$$

Clearly, φ is compact valued. We shall show that it is upper semicontinuous when Γ is endowed with the product topology and $C(K)$ is provided with the topology of pointwise convergence. So let $U \subset C(K)$ be a pointwise open set containing $\varphi(\sigma)$ for some $\sigma \in \Gamma$. As $0 \in U$, there are $A_1, \ldots, A_m \in \mathcal{A}$ and $\epsilon > 0$ such that $f \in U$ whenever $f \in C(K)$ and $|f(\chi_{A_i})| = |0(\chi_{A_i}) - f(\chi_{A_i})| < \epsilon$ for $i = 1, \ldots, m$. For every such i we find $n_i \in \mathbb{N}$ such that $\rho \notin A_i$ whenever $\rho \in \Gamma$, $\rho \neq \sigma$, and $\rho|n_i = \sigma|n_i$. Put $n = \max\{n_1, \ldots, n_m\}$. Now take any $\rho \in \Gamma$ different from σ and such that $\rho|n = \sigma|n$. Then $\rho \notin A_1 \cup \cdots \cup A_m$ and so $\pi_\rho \in U$. This proves that φ is usco. Thus, by Definition 4.1.1, the space $Y := \{\pi_\sigma : \sigma \in \Gamma\} \cup \{0\}$ is $\mathcal{K}$-analytic. Therefore (an analogue of) Theorem 7.1.8 guarantees that $C(K)$ is weakly $\mathcal{K}$-analytic.

     In order to prove that K is not an Eberlein compact, we need the following Theorem:

Theorem 4.3.2.

Let K be a compact subspace of $\{0,1\}^{\Gamma}$ for some set $\Gamma \neq \emptyset$ and assume that for every $k \in K$ there exists a sequence $\{y_n\} \subset K$, with finite supports supp y_n, such that $\lim_{n \to \infty} y_n = k$. Then K is Eberlein compact if and only if there exist sets $\Gamma_i \subset \Gamma$, $i = 1, 2, \ldots$, such that $\Gamma = \bigcup_{i=1}^{\infty} \Gamma_i$ and that for every $k = \chi_A \in K$ and every $i \in \mathbb{N}$ the set $A \cap \Gamma_i$ is finite.

Proof. Assume the condition is satisfied. We define a mapping $T : K \to [0,1]^\Gamma$ by $Tk(\gamma) = (1/i)k(\gamma)$ if $\gamma \in \Gamma_i \setminus \bigcup_{j=1}^{i-1} \Gamma_j$, $i = 1, 2, \dots$. Clearly, T is a one-to-one mapping into $c_0(\Gamma)$, and T is continuous when $c_0(\Gamma)$ is provided with the weak topology. Therefore, K is Eberlein compact; see Definition 1.2.1.

Now, assume that K is Eberlein compact. Let us observe an easily provable fact, valid for all totally disconnected spaces, that a base for the topology on K can be formed by the sets

$$W(M, L) = \{\chi_A \in K : M \subset A \subset \Gamma \setminus L\},$$

where M, L are any two finite (maybe empty) disjoint sets in Γ. Denote this family by $\mathcal{B}_0$. We also observe that each $W(M, L)$ is clopen, that is, simultaneously closed and open in K.

According to Theorem 1.2.4(i), (v), we find a family $\mathcal{U} = \bigcup_{n=1}^\infty \mathcal{U}_n$ of open F_σ sets such that $\mathcal{U}$ separates the points of K (i.e., whenever $k, h \in K$ are different, then $\{k, h\} \cap U$ is a singleton for some $U \in \mathcal{U}$) and each $\mathcal{U}_n$ is point finite (i.e., for every $k \in K$ the family $\{U \in \mathcal{U}_n : U \ni k\}$ is finite).

In what follows, we shall construct another such family which has the extra property that every element of it is a clopen set. To do this, fix any $n \in \mathbb{N}$ and any $U \in \mathcal{U}_n$. As U is F_σ, we may write $U = \bigcup_{i=1}^\infty F_i$, where each F_i is a closed set. From compactness we know that every F_i can be covered by finitely many elements from the basis $\mathcal{B}_0$ which are subsets of U. Thus, we can conclude that there are sets $W_1^U, W_2^U, \dots$ in $\mathcal{B}_0$ such that $U = \bigcup_{i=1}^\infty W_i^U$. Now define $\mathcal{V} = \bigcup_{n=1}^\infty \mathcal{V}_n$, where $\mathcal{V}_n = \{W_i^U : U \in \mathcal{U}_m, \, i, m \in \{1, \dots, n\}\}$. We can easily check that the family $\mathcal{V}$ separates the points of K, that every $\mathcal{V}_n$ is point finite, and, of course, the elements of $\mathcal{V}$ are clopen sets.

Put $\mathcal{B}_1 = \mathcal{V} \cup \{K \setminus U : U \in \mathcal{V}\}$. We claim that $\mathcal{B}_1$ *is a subbase for the topology of K.* So fix any open set $\emptyset \neq \Omega \subset K$ and any $x \in \Omega$. For every $y \in K \setminus \Omega$ we find $U_y \in \mathcal{V}$ such that $\{x, y\} \cap U_y$ is a singleton. From compactness, we select $y_1, \dots, y_j, y_{j+1}, \dots, y_m \in K \setminus \Omega$ such that $y_i \notin U_{y_i}$ for $i \in \{1, \dots, j\}$, that $y_i \in U_{y_i}$ for $i \in \{j+1, \dots, m\}$, and that

$$K \setminus \Omega \subset \bigcup_{i=1}^{j} (K \setminus U_{y_i}) \cup \bigcup_{i=j+1}^{m} U_{y_i}.$$

. Then

$$\Omega \supset \bigcap_{i=1}^{j} U_{y_i} \cap \bigcap_{i=j+1}^{m} (K \setminus U_{y_i}) \; (\ni x).$$

This proves the claim. Let $\mathcal{B}_2$ be the base constructed from $\mathcal{B}_1$, that is, $\mathcal{B}_2 = \{B_1 \cap \cdots \cap B_m : B_1, \dots, B_m \in \mathcal{B}_1, \, m \in \mathbb{N}\}$.

Now put $S_1 = \{\chi_U, \chi_{K \setminus U} : U \in \mathcal{V}_1\} \cup \{1\} \cup \{0\}$, where 1 and 0 mean the functions on K equal identically to 1 and 0, respectively. Since $\mathcal{V}_1$ is point finite, S_1 is a pointwise compact set in the space $C(K, \{0, 1\})$ of continuous functions from K to $\{0, 1\}$. If $i \in \mathbb{N}$ and S_i has already been defined, then we put

$$S_{i+1} = \{\min(f, g), \ \max(f, g) : f, g \in S_i\} \cup \{\chi_U, \chi_{K \setminus U} : U \in \mathcal{V}_{i+1}\}.$$

By induction, we can easily verify that every S_i is a pointwise compact set. We can also easily check that $\chi_\Omega \in \bigcup_{i=1}^\infty S_i$ for every $\Omega \in \mathcal{B}_2$. Now, consider any ϕ from $C(K, \{0, 1\})$. Then $\phi = \chi_A$, where A is a clopen set in K. Hence, from compactness, there are $\Omega_1, \ldots, \Omega_m \in \mathcal{B}_2$ such that $A = \Omega_1 \cup \cdots \cup \Omega_m$. Thus, $\phi = \max(\chi_{\Omega_1}, \ldots, \chi_{\Omega_m})$ and so ϕ belongs to S_i for some large $i \in \mathbb{N}$. In this way, we have shown that $C(K, \{0, 1\}) = \bigcup_{i=1}^\infty S_i$, which is a σ-compact space in the pointwise topology.

For $\gamma \in \Gamma$ we define $\pi_\gamma \in C(K, \{0, 1\})$ by $\pi_\gamma(k) = k(\gamma)$, $k \in K$. Now, for $i = 1, 2, \ldots$ we put $\Gamma_i = \{\gamma \in \Gamma : \pi_\gamma \in S_i\}$. We shall show that Γ_i are the desired sets. So take any $\chi_A \in K$ and any $i \in \mathbb{N}$ and assume that $A \cap \Gamma_i$ is an infinite set. Take an infinite one-to-one sequence $\{\gamma_n\}$ in it. We observe that for every finite set $F \subset \Gamma$, with $\chi_F \in K$, we have

$$\lim_{n \to \infty} \pi_{\gamma_n}(\chi_F) = \lim_{n \to \infty} \chi_F(\gamma_n) = 0.$$

Since S_i is compact, the sequence $\{\pi_{\gamma_n}\}$ has a cluster point, say $\psi \in S_i$. Then $\psi(\chi_F) = 0$ for every $\chi_F \in K$ with finite $F \subset \Gamma$. But ψ is continuous. Thus, by the assumption, $\psi(\chi_A) = 0$. On the other hand, we have

$$\pi_{\gamma_n}(\chi_A) = \chi_A(\gamma_n) = 1$$

for all $n \in \mathbb{N}$ and hence $\psi(\chi_A) = 1$, a contradiction. Therefore the set $A \cap \Gamma_i$ is finite. $\qquad\Box$

Back to the Proof of Theorem 4.3.1. Assume that the K in this theorem is Eberlein compact. The assumption of Theorem 4.3.2 can be easily verified. So $\mathbb{N}^{\mathbb{N}}$ can be written as the union of sets Γ_i, $i = 1, 2, \ldots$, such that for every $k = \chi_A \in K$ and every $i \in \mathbb{N}$ the set $A \cap \Gamma_i$ is finite. However, according to Lemma 1.6.1, this is not the case.

Assume now that $C(K)$ is isomorphic to a subspace of a WCG space V. Then, by Theorems 1.2.5, 1.2.4, (B_{V^*}, w^*) is Eberlein compact and a simple application of the Hahn–Banach theorem reveals that $(B_{C(K)^*}, w^*)$ is a subspace of a continuous image of (B_{V^*}, w^*). Hence, according to a theorem of Benyamini, M. E. Rudin, and Wage [BRW], $(B_{C(K)^*}, w^*)$ is Eberlein compact and so K is also Eberlein compact, a contradiction.

That $C(K)$ is not even isomorphic to a subspace of an Asplund generated space will be guaranteed by Theorem 8.3.7. $\qquad\Box$

4.4. NOTES AND REMARKS

Theorem 4.1.2 is due to Debs [De]. In its proof we freely followed his argument with two exceptions. We avoided working with Souslin operation and we did not use the fact that the space of continuous functions on a metrizable compact is separable in the supremum norm [En, Corollary 4.2.18]. We also note that Theorem 4.1.2 can be deduced from Theorem 4.2.3. In fact, according to Mercourakis [Me], we know that the dual of a weakly $\mathcal{K}$-analytic space, even of a Vašák space, admits a strictly convex dual norm. Indeed, this follows from Theorem 7.2.5 and the fact that the space $c_1(\Sigma' \times \Gamma)$ is strictly convexifiable; see [Me].

It does not seem that Debs's proof would work for Vašák spaces, that is, for those Banach spaces V for which $V = \varphi(\Sigma')$, where $\Sigma' \subset \mathbb{IN}^{\mathbb{IN}}$ and φ is a usco mapping from Σ' onto (V, w). Indeed, the completeness of the space $\mathbb{IN}^{\mathbb{IN}}$ probably plays an essential role in the proof of Theorem 4.1.2. Yet, using another argument, we shall obtain in Chapter 7 that the Vašák spaces do belong to $\widetilde{S}$, and even to the class $\widetilde{\mathcal{F}}$, and so are weak Asplund; see Theorem 7.2.8.

Lemma 4.2.1 is from Oxtoby's book [Ox]. In the proof of the sufficiency part, we considered a more complicated strategy than that found in the necessity part. This was because, in practice, when we do not know a priori that M is a residual set, it is difficult to find the function f. It should be noted that, using Lemma 4.2.1, we get that *the image of a residual set in a complete metric space under a continuous open surjective mapping is residual.* In this way, we can reprove Theorem 2.1.5; see [Ph, Lemma 4.25].

Theorem 4.2.3, with a different proof, first appeared in Jokl [Jo]. If we replace coFz in the conclusion of Proposition 4.2.2 by Fz, then there exists a more direct proof of it. In fact, ideas of Kenderov [K1] can be adapted a little: Put $f(z) = \min\{\|\xi\| : \xi \in Fz\}$, $z \in Z$. We can easily verify that f is a lower semicontinuous function. Hence there is a residual set M in Z such that f is continuous at each point of M [En, 1.7.14]. Fix one $z \in M$. We claim that $\|\xi\| = f(z)$ for each $\xi \in Fz$. Assume not. Then there are $\epsilon > 0$ and $\xi \in Fz$ such that $\|\xi\| > f(z) + \epsilon$. Find $e \in V$, $\|e\| = 1$, satisfying $\langle \xi, e \rangle > f(z) + \epsilon$. Putting $W = \{v^* \in V^* : \langle v^*, e \rangle > f(z) + \epsilon\}$, we have $Fz \cap W \neq \emptyset$. Now, from the continuity of f we can find an open set $\Omega \ni z$ such that $f(x) < f(z) + \epsilon$ whenever $x \in \Omega$. On the other hand, remarking that $F(\Omega) \cap W \neq \emptyset$ and using the minimality of F, Lemma 3.1.2(iii) yields a nonempty open set $U \subset \Omega$ such that $F(U) \subset W$ and so $f(x) \geq f(z) + \epsilon$ for all $x \in U$, which is clearly false. Therefore $\|\xi\| = f(z)$ for each ξ in Fz.

Theorem 4.2.4 is from a paper by Preiss, Phelps, and Namioka [PPN]. It is a real extension of Theorem 4.2.3 because there exists a Gâteaux smooth, non-Asplund space whose dual admits no equivalent strictly convex dual norm. Indeed, $C([0, \omega_1]) \times \ell_1$ is such a space; see Talagrand's Fréchet smooth renorming of $C([0, \omega_1])$ [Ta5] and [DGZ2, Theorem II.2.6(i)]. Theorem 4.2.4 will be strengthened in the next chapter; see Theorem 5.3.1. The method of subsequent changing of norm used in the proof of Theorem 4.2.4 is freely transferred from Preiss [Pr], where it worked in a different situation. A further easy cultivation of this technique yields the result: *If V is a Banach space provided with a bornology β and admits a β-smooth norm, then every minimal usco mapping from a Baire space into (V^*, w^*) is β-continuous at the points of a residual set;* see [PPN]. Thus, by an appropriate choice of the family β, we get Theorem 4.2.4 as well as the well-known fact that *Banach spaces with Fréchet smooth norm are Asplund;* see [Ph, Corollary 4.15] and [EL].

Corollary 4.2.5 strengthens a result of Borwein and Preiss that *every convex continuous function on a Gâteaux smooth Banach space is Gâteaux differentiable at the points of a dense set* [BP]. Yet some serious questions remain untouched: *Does every weak Asplund space belong to class $\tilde{S}$?* Or, more modestly: *Is every monotone mapping on a weak Asplund space single-valued at the points of a residual set?*

In Section 4.3, we followed Talagrand [Ta2]. The space $C(K)$ constructed there is the first example in the literature of a weakly $\mathcal{K}$-analytic Banach space which is not a subspace of a WCG space. Actually, more can be said: $C(K)$ *is a $K_{\sigma\delta}$ set in $(C(K)^{**}, w^*)$, that is, there are compact sets $K_{np} \subset (C(K)^{**}, w^*)$ such that $C(K) = \bigcap_{p=1}^{\infty} \bigcup_{n=1}^{\infty} K_{np}$* [Ta2]. (It is easy to check that *a subspace of any WCG space is $K_{\sigma\delta}$ in its second dual with weak* topology and that this implies weak $\mathcal{K}$-analyticity.*) So far it is unclear *if every weakly $\mathcal{K}$-analytic Banach space V is $K_{\sigma\delta}$ in (V^{**}, w^*).* A dual weakly $\mathcal{K}$-analytic space which is not isomorphic to a subspace of a WCG space was constructed by Mercourakis [AM]; he used ideas of Kutzarova and Troyanski [KT].

In [OSV], it is shown that the compact K from Section 4.3 does not admit a lower semicontinuous metric that fragments it. Hence K is not a Radon–Nikodým compact by [N1, N2]. Therefore, Theorem 8.3.5 does not allow K to be Eberlein compact.

Another important example of a Talagrand compact, that is, a compact K such that $C(K)$ is weakly $\mathcal{K}$-analytic, was constructed by E. A. Rezničenko; see Section 8.4 [Arg, K5]. This compact is not Eberlein according to Lemma 8.4.2 or because it does not admit a lower semicontinuous fragmenting metric [K5].

There is a rich theory of weakly $\mathcal{K}$-analytic spaces; see [Ta2, So, Me]. However, we did not pay much attention to it, because there exists a slightly larger class: Vašák spaces, which have a very similar theory; see Chapter 7.

There is a generalization of Theorem 4.3.2 due to Farmaki [Fa]: *A compact $K \subset \Sigma(\Gamma)$ is Eberlein if and only if for every $\epsilon > 0$ there exist sets $\Gamma_n^\epsilon \subset \Gamma$, $n = 1, 2, \ldots,$ with $\bigcup_{n=1}^{\infty} \Gamma_n^\epsilon = \Gamma$, and such that for every $k \in K$ and every $n \in \mathbb{N}$ the set $\{\gamma \in \Gamma_n^\epsilon : |k(\gamma)| > \epsilon\}$ is finite.* There also exist characterizations of uniform Eberlein compacta and of Talagrand compacta, which are in this vein; see [Fa].

Chapter Five

Fragmentability

The main concept we shall deal with in this chapter is that of fragmentability — a natural generalization of dentability. It will prove to be useful in ensuring weak Asplundness of some quite significant classes of Banach spaces.

Definition 5.0.1. Let X be a topological space and ρ be a mertric on X. We say that X is *fragmented* by ρ or that ρ *fragments* X if for every $\epsilon > 0$ and every nonempty subset M of X there exists an open subset $\Omega \subset X$ such that the set $M \cap \Omega$ is nonempty and has ρ-diameter less than ϵ. The space X is said to be *fragmentable* if it is fragmented by some metric on it. The class of all fragmentable topological spaces is denoted by $\mathcal{F}$.

Of course, every metric space is fragmentable. Thus, in particular, bounded sets with weak* topology in the dual of a separable Banach space are fragmentable. Bounded sets with weak* topology in the dual of an Asplund space are also fragmentable; see Theorem 1.1.1. Hence, Radon–Nikodým, in particular Eberlein compacta are fragmentable. In all these spaces, the fragmenting metric is lower semicontinuous. For fragmentable spaces whose fragmenting metric is not necessarily lower semicontinuous, see Sections 5.3 and 7.2.

In Section 5.1, we first present a purely topological and simply formulated characterization of fragmentability. Then we use it to show that $\mathcal{F}$ has several nice properties and that it is a subfamily of Stegall's class $\mathcal{S}$. Section 5.2 deals with a Banach space counterpart $\widetilde{\mathcal{F}}$ of $\mathcal{F}$. We prove that if a compact space K is fragmentable, then $C(K)$ belongs to $\widetilde{\mathcal{F}}$. Section 5.3 is devoted to the proof of a delicate fact that Banach spaces with Gâteaux smooth norms belong to $\widetilde{\mathcal{F}}$, thus strenghtening Theorem 4.2.4.

5.1. CLASS $\mathcal{F}$ OF FRAGMENTABLE TOPOLOGICAL SPACES

We shall first show that fragmentability is equivalent to a simple covering property of the space.

Definition 5.1.1. A well-ordered family $\mathcal{U} = \{U_\xi : \xi \in [0, \xi_0)\}$ of subsets of a topological space X is said to be a *relatively open partitioning* of X if

(i) U_ξ is contained and is relatively open in $X \setminus \left(\bigcup_{\eta < \xi} U_\eta \right)$ for every $\xi \in [0, \xi_0)$; and

(ii) $X = \bigcup_{\xi < \xi_0} U_\xi$.

Definition 5.1.2. Let $\mathcal{W} = \{W_\xi : \xi \in (0, \xi_0]\}$ be a family of subsets of a topological space X. We say that $\mathcal{W}$ is *regularly increasing* if

(i) $W_\xi \subset W_\eta$ whenever $0 < \xi < \eta \leq \xi_0$;

(ii) $W_\xi = \bigcup_{\eta < \xi} W_\eta$ for every limit ordinal $\xi \in (0, \xi_0]$; and

(iii) $W_{\xi_0} = X$.

The following proposition relates these two concepts to each other:

Proposition 5.1.3.

Let $\mathcal{U} = \{U_\xi : \xi \in [0, \xi_0)\}$ be a relatively open partitioning of a topological space X and let us put $W_\xi = \bigcup_{\eta < \xi} U_\eta$, $\xi \in (0, \xi_0]$. Then $\mathcal{W}(\mathcal{U}) := \{W_\xi : \xi \in (0, \xi_0]\}$ is a regularly increasing family of open subsets of X and $U_\xi = W_{\xi+1} \setminus W_\xi$ for every $\xi \in (0, \xi_0)$.

Let a family $\mathcal{W} = \{W_\xi : \xi \in (0, \xi_0]\}$ of open subsets of a topological space X be regularly increasing. Then the family $\mathcal{U} = \{U_\xi : \xi \in [0, \xi_0)\}$, where $U_0 = W_1$ and $U_\xi = W_{\xi+1} \setminus W_\xi$, $\xi \in (0, \xi_0)$, is a relatively open partitioning of X and $W_\xi = \bigcup_{\eta < \xi} U_\eta$ for every $\xi \in (0, \xi_0]$.

Proof. First part. Conditions (i) – (iii) from Definition 5.1.2 can be verified immediately. Also $U_\xi = W_{\xi+1} \setminus W_\xi$ surely holds for each $\xi \in (0, \xi_0)$. It remains to prove that each W_ξ is open. We shall proceed by transfinite induction. The set W_1 is open since it is equal to U_0. Assume that each W_η is open whenever $\eta \in (0, \xi)$, where ξ is a fixed element of $(1, \xi_0]$. If ξ is a limit ordinal, then, clearly, W_ξ is open. If ξ is not a limit ordinal, then we can write $\xi = \zeta + 1$ and

$$U_\zeta = \Omega \cap \left(X \setminus \bigcup_{\eta < \zeta} U_\eta \right)$$

with some open set $\Omega \subset X$; so

$$W_\xi = \bigcup_{\eta < \xi} U_\eta = W_\zeta \cup U_\zeta = W_\zeta \cup (\Omega \setminus W_\zeta) = W_\zeta \cup \Omega$$

is open. This completes the induction step, and the first part of the proof of the proposition is finished.

The second part can be proved even more easily.   □

Definition 5.1.4. Let $\mathcal{U}, \mathcal{V}$ be two relatively open partitionings of a topological space X. We say that $\mathcal{V}$ is a *refinement* of $\mathcal{U}$ if the regularly increasing family $\mathcal{W}(\mathcal{U})$ corresponding to $\mathcal{U}$ by Proposition 5.1.3 is contained in the regularly increasing family $\mathcal{W}(\mathcal{V})$ corresponding to $\mathcal{V}$. We say that $\mathcal{V}$ is a *strong refinement* of $\mathcal{U}$ if $\mathcal{V}$ is a refinement of $\mathcal{U}$ and for every $V \in \mathcal{V}$ there is $U \in \mathcal{U}$ such that $\bar{V} \subset U$.

Proposition 5.1.5.

Let $\mathcal{U} = \{U_\xi : \xi \in [0, \xi_0)\}$ be a relatively open partitioning of X and for every $\xi \in [0, \xi_0)$ let $\mathcal{U}^\xi = \{U_\eta^\xi Z : \eta \in [0, \eta_\xi)\}$ be a relatively open partitioning of U_ξ with the topology inherited from X. Let us order the set

$$I = \{(\xi, \eta) : \xi \in [0, \xi_0), \ \eta \in [0, \eta_\xi)\}$$

lexicographically, that is, $(\xi_1, \eta_1) < (\xi_2, \eta_2)$ if and only if either $\xi_1 < \xi_2$ or $\xi_1 = \xi_2$ and $\eta_1 < \eta_2$. Then the family $\mathcal{V} = \{U_\eta^\xi : (\xi, \eta) \in I\}$ is a relatively open partitioning of X, which refines $\mathcal{U}$.

Proof. Condition (ii) from Definiton 5.1.1 is easily verified. We shall prove (i). Take any $(\xi, \eta) \in I$. There is an open set $\Omega \subset X$ so that

$$U_\xi = \Omega \backslash \bigcup_{\xi' < \xi} U_{\xi'}.$$

Further, there is an open set $G \subset X$ so that

$$U_\eta^\xi = (G \cap U_\xi) \backslash \bigcup_{\eta' < \eta} U_{\eta'}^\xi.$$

Then

$$U_\eta^\xi = \left(G \cap \left(\Omega \backslash \bigcup_{\xi' < \xi} U_{\xi'} \right) \right) \backslash \bigcup_{\eta' < \eta} U_{\eta'}^\xi = (G \cap \Omega) \backslash \left(\bigcup_{\xi' < \xi} U_{\xi'} \cup \bigcup_{\eta' < \eta} U_{\eta'}^\xi \right)$$

$$= (G \cap \Omega) \backslash \bigcup \{ U_{\eta'}^{\xi'} : (\xi', \eta') \in I, \ (\xi', \eta') < (\xi, \eta) \},$$

which verifies (i) of Definition 5.1.1. We have thus shown that $\mathcal{V}$ is a relatively open partitioning of X. Moreover, $\mathcal{V}$ is a refinement of $\mathcal{U}$ because for every $\xi \in (0, \xi_0)$

$$\bigcup_{\xi' < \xi} U_{\xi'} = \bigcup \{ U_{\eta'}^{\xi'} : (\xi', \eta') \in I, \ (\xi', \eta') < (\xi, 0) \}.$$

$\square$

Proposition 5.1.6.

Let $\mathcal{U}$ be a relatively open partitioning of a regular space X. Then there exists on X a relatively open partitioning $\mathcal{V}$ which is a strong refinement of $\mathcal{U}$.

Proof. Write $\mathcal{U} = \{U_\xi : \xi \in [0, \xi_0)\}$. The regularity of X gives, for <u>every</u> $\xi \in (0, \xi_0)$ and every $x \in U_\xi$, an open set $V_x^\xi \subset X$ such that $x \in V_x^\xi$ and $\overline{V_x^\xi} \subset \bigcup_{\eta \leq \xi} U_\eta$. Let us fix any $\xi \in [0, \xi_0)$ and let us well order the set U_ξ, that is, $U_\xi = \{x_\eta : \eta \in [0, \eta_\xi)\}$, where η_ξ is an ordinal [En, Section I.4]. Then $\mathcal{U}^\xi = \{U_\eta^\xi : \eta \in [0, \eta_\xi)\}$, where

$$U_\eta^\xi = V_{x_\eta}^\xi \setminus \left(\bigcup_{\zeta < \xi} U_\zeta \cup \bigcup_{\zeta < \eta} V_{x_\zeta}^\xi \right)$$

is a relatively open partitioning of U_ξ, for which

$$\overline{U_\eta^\xi} \subset \overline{V_{x_\eta}^\xi \setminus \bigcup_{\zeta < \xi} U_\zeta} \subset \overline{V_{x_\eta}^\xi} \bigcap \left(X \setminus \bigcup_{\zeta < \xi} U_\zeta \right) \subset \bigcup_{\zeta \leq \xi} U_\zeta \setminus \bigcup_{\zeta < \xi} U_\zeta = U_\xi.$$

Proposition 5.1.5 now completes the proof.   □

Definition 5.1.7. Let $\mathcal{U}$ be a family of subsets of a topological space X. We say that $\mathcal{U}$ *separates the points* of X if for every two distinct points $x, y \in X$ there is $U \in \mathcal{U}$ such that $\{x, y\} \cap U$ is a singleton. The family $\mathcal{U}$ is said to be a *σ-relatively open partitioning* of X if there are relatively open partitionings $\mathcal{U}_n$ of X, $n = 1, 2, \ldots$, such that $\mathcal{U} = \bigcup_{n=1}^{\infty} \mathcal{U}_n$. The family $\mathcal{U}$ is called a *separating σ-relatively open partitioning* of X if it separates the points of X and is a σ-relatively open partitioning of X.

It is easy to check that if a σ-relatively open partitioning $\mathcal{U} = \bigcup_{n=1}^{\infty} \mathcal{U}_n$ separates the points, then so does the union of the corresponding regularly increasing families $\mathcal{W}(\mathcal{U}_n)$ and vice versa.

Proposition 5.1.8.

Suppose that a (regular) topological space X admits a separating σ-relatively open partitioning. Then there exist relatively open partitionings $\mathcal{U}_n$ of X, $n = 1, 2, \ldots$, such that $\mathcal{U}_{n+1}$ is a (strong) refinement of $\mathcal{U}_n$, for each n, and $\bigcup_{n=1}^{\infty} \mathcal{U}_n$ separates the points of X.

Proof. Let $\mathcal{V} = \bigcup_{n=1}^{\infty} \mathcal{V}_n$ be a separating σ-relatively open partitioning of X. We set $\mathcal{U}_1 = \mathcal{V}_1$. Suppose that we have already constructed $\mathcal{U}_1, \mathcal{U}_2, \ldots, \mathcal{U}_n$ for some $n \in \mathbb{N}$ such that $\mathcal{U}_{i+1}$ is a (strong) refinement of $\mathcal{U}_i$ for $i = 1, \ldots, n-1$. Write

$$\mathcal{U}_n = \{U_\xi : \xi \in [0, \xi_0)\}, \qquad \mathcal{V}_{n+1} = \{V_\eta : \eta \in [0, \eta_0)\}$$

and endow the set $[0, \xi_0) \times [0, \eta_0)$ with the lexicographical order. Then the family $\tilde{\mathcal{U}}_{n+1} = \{U_\xi \cap V_\eta : (\xi, \eta) \in [0, \xi_0) \times [0, \eta_0)\}$ is, by Proposition 5.1.5, a relatively open partitioning of X, which is a refinement of $\mathcal{U}_n$. In the nonparenthetic case simply put $\mathcal{U}_{n+1} = \tilde{\mathcal{U}}_{n+1}$. If X is regular, then Proposition

5.1.6 yields a relatively open partitioning of X, say $\mathcal{U}_{n+1}$, which is a strong refinement of $\tilde{\mathcal{U}}_{n+1}$; hence a fortiori it is a strong refinement of $\mathcal{U}_n$. Then, surely, $\bigcup_{n=1}^{\infty} \mathcal{U}_n$ separates the points of X. $\qquad\square$

Now we are ready to present the topological characterization of fragmentability.

Theorem 5.1.9.

A topological space is fragmentable if and only if it admits a separating σ-relatively open partitioning.

Proof. Let a topological space X admit a separating σ-relatively open partitioning $\mathcal{U} = \bigcup_{n=1}^{\infty} \mathcal{U}_n$ with

$$\mathcal{U}_n = \left\{ U_\xi^n : \xi \in [0, \xi_n) \right\}.$$

According to Proposition 5.1.8, we may and do assume that $\mathcal{U}_{n+1}$ is a refinement of $\mathcal{U}_n$ for every $n \in \mathbb{N}$. We define a mapping $\rho : X \times X \to [0, +\infty)$ by

$$\rho(x, y) = \begin{cases} 0 & \text{if } x = y, \\ \left(\min \left\{ n \in \mathbb{N} : \#(U \cap \{x, y\}) = 1 \text{ for some } U \in \mathcal{U}_n \right\} \right)^{-1} & \text{if } x \neq y. \end{cases}$$

It is not difficult to check that ρ is a metric on X. In fact, the triangle inequality can be replaced by a stronger one:

$$\rho(x, y) \le \max \left\{ \rho(x, z), \rho(z, y) \right\}, \qquad x, y, z \in X.$$

Indeed, take $x, y, z \in X$, $x \neq y$, and denote $n = \rho(x, y)^{-1}$. We find $U \in \mathcal{U}_n$ so that $\{x, y\} \cap U$ is a singleton, say $x \in U$ and $y \notin U$. If $z \in U$, then $\{z, y\} \cap U$ is a singleton and so $\rho(z, y) \ge 1/n$ $(= \rho(x, y))$. If $z \notin U$, then $\{x, z\} \cap U$ is a singleton and hence $\rho(x, z) \ge 1/n$ $(= \rho(x, y))$.

Moreover, the ρ-diameter of U_ξ^n is less than $1/n$ for every $n \in \mathbb{N}$ and every $\xi \in [0, \xi_n)$ since $\mathcal{U}_{i+1}$ is a refinement of $\mathcal{U}_i$. We shall show that X is fragmented by ρ. So let $\epsilon > 0$ and an arbitrary set $\emptyset \neq Y \subset X$ be given. Take $n > 1/\epsilon$ and put.

$$\zeta = \min \left\{ \eta \in [0, \xi_n) : Y \cap U_\eta^n \neq \emptyset \right\}.$$

Then the (nonempty) set $Y \cap U_\zeta^n$ is equal to $Y \cap \left(\bigcup_{\eta \le \zeta} U_\eta^n \right)$ and hence is open in Y. Moreover,

$$\rho\text{-diam} \left(Y \cap U_\zeta^n \right) \le \rho\text{-diam } U_\zeta^n < 1/n < \epsilon.$$

We have proved that ρ fragments X.

Now assume that X is fragmented by a metric ρ. For every $n \in \mathbb{N}$ we shall construct a relatively open partitioning $\mathcal{U}_n$ of X as follows. Fix any $n \in \mathbb{N}$. We shall proceed by transfinite induction. Set $U_0 = \emptyset$. Let $\xi > 0$ be a given ordinal and let us assume that for every $\zeta \in [0, \xi)$ we have found a set $U_\zeta \subset X$ which lies in $X \setminus \bigcup_{\eta < \zeta} U_\eta$ and is relatively open there and with ρ-diam $U_\zeta < 1/n$. If $\bigcup_{\eta < \xi} U_\eta = X$, then we finish the process and put $\mathcal{U}_n = \{ U_\eta : \eta \in [0, \xi) \}$. In what follows, let us assume that this is not the case. Then there exists, by assumption, a nonempty relatively open subset U_ξ in $X \setminus \bigcup_{\eta < \xi} U_\eta$ with ρ-diameter less than $1/n$. This process must eventually stop. Clearly, $\mathcal{U}_n$ is then a relatively open partitioning of X. It remains to show that $\bigcup_{n=1}^{\infty} \mathcal{U}_n$ separates the points of X. Let $x, y \in X$, $x \neq y$. There is $n \in \mathbb{N}$ so that $\rho(x, y) \geq 1/n$. Further there is $U \in \mathcal{U}_n$ such that $x \in U$ and, as we know that ρ-diam $U < 1/n$, y cannot belong to U. $\square$

Let us remark that we could consider a topological space which is "fragmented" by any function $\tau : X \times X \to [0, +\infty)$ with the property that $\tau(x, y) = 0$ if and only if $x = y$. The above proof then shows that X with such a τ is fragmented by a metric.

In what follows, we shall show a complete analogue of Theorem 3.1.5. This will have useful consequences for Banach spaces.

Theorem 5.1.10.

(i) If $X \in \mathcal{F}$ and $g : X \to Y$ is a surjective perfect mapping, that is, g is continuous and maps closed sets onto closed sets and $g^{-1}(y)$ is a compact set for every $y \in Y$, then $Y \in \mathcal{F}$; in particular, a continuous image of a fragmentable compact is fragmentable.

(ii) If $Y \in \mathcal{F}$ and X continuously injects into Y, then $X \in \mathcal{F}$; in particular, every subspace of Y belongs to $\mathcal{F}$.

(iii) If $X_i \in \mathcal{F}$, $i = 1, 2, \ldots$, then $\prod_{i=1}^{\infty} X_i \in \mathcal{F}$.

(iv) If $X = \bigcup_{i=1}^{\infty} X_i$, where all X_i are closed in X and belong to $\mathcal{F}$, then $X \in \mathcal{F}$.

Proof. (ii) Let Y be fragmented by a metric ρ and let $f : X \to Y$ be a continuous injection. Then it is straightforward to check that $(x_1, x_2) \mapsto \rho(f(x_1), f(x_2))$ is a metric which fragments X.

(iii) Let $X_i \in \mathcal{F}$, $i \in \mathbb{N}$. By Theorem 5.1.9, each X_i has a separating σ-relatively open partitioning $\bigcup_{n=1}^{\infty} \mathcal{U}_n^i$ where

$$\mathcal{U}_n^i = \left\{ U_\xi^{in} : \xi \in [0, \xi_n^i) \right\}.$$

Define

$$\mathcal{V}_{in} = \left\{ U_\xi^{in} \times \prod_{\substack{j=1 \\ j \neq i}}^{\infty} X_j : \xi \in [0, \xi_n^i) \right\}, \qquad i, n \in \mathbb{N}.$$

Then $\bigcup \{ \mathcal{V}_{in} : i, n \in \mathbb{N} \}$ is clearly a separating σ-relatively open partitioning of $\prod_{j=1}^{\infty} X_j$ and Theorem 5.1.9 completes the proof of (iii).

(iv) We again use the characterization given in Theorem 5.1.9. For $i = 1, 2, \ldots$ let $\bigcup_{n=1}^{\infty} \mathcal{U}_n^i$ be a separating σ-relatively open partitioning of X_i, where

$$\mathcal{U}_n^i = \left\{ U_\xi^{in} : \xi \in [0, \xi_n^i) \right\}.$$

For $i, n \in \mathbb{N}$ we define $\mathcal{V}_n^i = \{ V_\xi^{in} : \xi \in [-1, \xi_n^i) \}$, where

$$V_{-1}^{in} = X \backslash X_i \qquad \text{and} \qquad V_\xi^{in} = U_\xi^{in}, \qquad \xi \in [0, \xi_n^i).$$

Then, clearly, $\bigcup \{ \mathcal{V}_n^i : i, n \in \mathbb{N} \}$ is a separating σ-relatively open partitioning of the whole space X.

The proof of (i) needs more attention. Let $\mathcal{U} = \bigcup_{n=1}^{\infty} \mathcal{U}_n$ be a separating σ-relatively open partitioning of X. By Proposition 5.1.8, we may assume that $\mathcal{U}_{n+1}$ is a refinement of $\mathcal{U}_n$ for each $n \in \mathbb{N}$. As usual, write

$$\mathcal{U}_n = \left\{ U_\xi^n : \xi \in [0, \xi_n) \right\} \qquad \text{and} \qquad W_\xi^n = \bigcup_{\eta < \xi} U_\eta^n, \qquad \xi \in (0, \xi_n], \qquad n \in \mathbb{N}.$$

For $n \in \mathbb{N}$ we define

$$\widetilde{\mathcal{W}}_n = \left\{ \widetilde{W}_\xi^n : \xi \in (0, \xi_n] \right\}, \qquad \text{where} \qquad \widetilde{W}_\xi^n = Y \backslash g(X \backslash W_\xi^n), \qquad \xi \in (0, \xi_n].$$

The $\widetilde{\mathcal{W}}_n$ are regularly increasing families of open subsets in Y. Indeed, since g sends closed sets to closed sets, every $\widetilde{W}_\xi^n$ is open. Further, if $\xi \in (0, \xi_n]$ is a limit ordinal, then

$$\widetilde{W}_\xi^n = Y \backslash g(X \backslash W_\xi^n) = Y \backslash g\left(\bigcap_{\eta < \xi} (X \backslash W_\eta^n) \right) = Y \backslash \bigcap_{\eta < \xi} g(X \backslash W_\eta^n) = \bigcup_{\eta < \xi} \widetilde{W}_\eta^n$$

because of the compactness of $g^{-1}(y)$ for every $y \in Y$. Let $\mathcal{V}_n$ be the relatively open partitioning of Y corresponding to $\widetilde{W}_\xi^n$ by Proposition 5.1.3.

It remains to prove that $\mathcal{V} := \bigcup_{n=1}^{\infty} \mathcal{V}_n$ separates the points of Y. Let y_1, y_2 be two distinct points of Y. For $n \in \mathbb{N}$ denote

$$\xi_i^n = \min \left\{ \xi \in (0, \xi_n] : g^{-1}(y_i) \subset W_\xi^n \right\}$$
$$\left(= \min \left\{ \xi \in (0, \xi_n] : y_i \in \widetilde{W}_\xi^n \right\} \right), \qquad i = 1, 2.$$

Because the sets $g^{-1}(y_i)$ are compact, ξ_i^n are not limit ordinals; hence $\xi_i^n - 1$ exist. We shall show that $\{g^{-1}(y_i)\backslash W_{\xi_i^n-1}^n\}_{n=1}^{\infty}$, $i = 1, 2$ are two nonincreasing sequences. So fix $n \in \mathrm{IN}$ and $i \in \{1, 2\}$. Because $\mathcal{U}_{n+1}$ is a refinement of $\mathcal{U}_n$, we have $W_{\xi_i^n-1}^n = W_{\xi}^{n+1}$ for some $\xi \in (0, \xi_{n+1}]$. Thus, $g^{-1}(y_i)\backslash W_{\xi}^{n+1} \neq \emptyset$ and hence $\xi < \xi_i^{n+1}$, that is, $\xi \leq \xi_i^{n+1} - 1$. Therefore $W_{\xi_i^n-1}^n \subset W_{\xi_i^{n+1}-1}^{n+1}$. It follows, using compactness, that there exists x_i in

$$\bigcap_{n=1}^{\infty} \left(g^{-1}(y_i)\backslash W_{\xi_i^n-1}^n\right), \qquad i = 1, 2.$$

The points x_1, x_2 are different, and hence there are $m \in \mathrm{IN}$ and $\eta \in (0, \xi_m)$ such that $\{x_1, x_2\} \cap W_{\eta}^m$ is a singleton, say $x_1 \in W_{\eta}^m$ and $x_2 \notin W_{\eta}^m$. It is clear then that $\eta < \xi_2^m$, that is, $\eta \leq \xi_2^m - 1$. Now if $\xi_1^m \geq \xi_2^m$, then we would have

$$x_1 \in g^{-1}(y_1)\backslash W_{\xi_1^m-1}^m \subset g^{-1}(y_1)\backslash W_{\eta}^m,$$

which is impossible. Therefore $\xi_1^m < \xi_2^m$. Thus, $g^{-1}(y_1) \subset W_{\xi_1^m}^m$ and $g^{-1}(y_2)\backslash W_{\xi_1^m}^m \neq \emptyset$. However, this means that $y_1 \in \widetilde{W}_{\xi_1^m}^m$ and $y_2 \notin \widetilde{W}_{\xi_1^m}^m$, that is, $\mathcal{V}$ separates the points y_1 and y_2. $\qquad\square$

Let us recall that a multivalued mapping $F : X \rightarrow 2^Y$ between two topological spaces X and Y is said to be *usco* if (i) it is upper semicontinuous, that is, for every open set $\Omega \subset Y$ the set $\{x \in X : Fx \subset \Omega\}$ is open, and (ii) for every $x \in X$ the set Fx is nonempty and compact. The mapping F is said to be *minimal usco* if it is usco and $F = G$ whenever $G : X \rightarrow 2^Y$ is usco and such that $Gx \subset Fx$ for each $x \in X$. The usefulness of fragmentability is demonstrated in the following basic result.

Theorem 5.1.11.

Let X, Y be two topological spaces, Y be fragmented by a metric ρ, and $F : X \rightarrow 2^Y$ be a minimal usco mapping. Then there exists a residual subset $\Omega \subset X$ such that for every $x \in \Omega$ the set Fx is a singleton and F is upper semicontinuous at x when Y is endowed with the topology generated by ρ. Hence, completely regular fragmentable spaces belong to Stegall's class $\mathcal{S}$.

Proof. For $n = 1, 2, \ldots$ we define sets

$$\Omega_n = \{x \in X : \text{there exists an open set } U \ni x \text{ with } \rho\text{-diam}\, F(U) < 1/n\}.$$

Clearly, Ω_n are open sets. They are also dense. Indeed, let U be an arbitrary open nonempty subset of X. From the fragmentability, there is an open set $G \subset Y$ such that $F(U) \cap G \neq \emptyset$ and $\rho\text{-diam}\left(F(U) \cap G\right) < 1/n$. Now, since F is minimal usco, there exists, by Lemma 3.1.2(iii), a nonempty open subset W of

U with $F(W) \subset G$. Then $F(W) \subset F(U) \cap G$ and hence W is contained in Ω_n. Thus, it remains to put $\Omega = \bigcap_{n=1}^{\infty} \Omega_n$. The last statement follows from Definition 3.1.3. $\qquad\qquad\square$

Theorem 5.1.12.

Let (K, τ) be a compact fragmentable space. Then K is sequentially compact and there exists a metric ρ on K such that
  (i) ρ fragments K;
  (ii) (K, ρ) is a complete metric space;
  (iii) ρ generates a topology stronger than τ; and
  (iv) there exists a dense G_δ (completely metrizable) set Ω in K such that τ coincides with the ρ-topology on Ω.

Proof. (i), (ii) By Theorem 5.1.9 and Proposition 5.1.8, there is a separating σ-relatively open partitioning $\mathcal{U} = \bigcup_{n=1}^{\infty} \mathcal{U}_n$ of K such that $\mathcal{U}_{n+1}$ is a strong refinement of $\mathcal{U}_n$ for each $n \in \mathbb{N}$. Then the metric ρ constructed in the proof of Theorem 5.1.9 from the family $\mathcal{U}$ is complete. In fact, let $\{k_i\}$ be a ρ-Cauchy sequence in K; so there exists an increasing sequence $\{i_m\} \subset \mathbb{N}$ such that $\rho(k_{i_m}, k_j) < 1/m$ if $j > i_m$. Take $U_m \in \mathcal{U}_m$ such that $k_{i_m} \in U_m$. Then, necessarily, $k_j \in U_m$ for all $j > i_m$. Indeed, $k_j \notin U_m$ would imply, by the definition of ρ, that $\rho(k_{i_m}, k_j) \geq 1/m$, a contradiction. Now note that for every $m \in \mathbb{N}$ we have $U_m \cap U_{m+1} \ni k_{i_{m+1}}$, so $U_{m+1} \subset U_m$ and hence $\overline{U_{m+1}} \subset U_m$ as $\mathcal{U}_{m+1}$ is a strong refinement of $\mathcal{U}_m$. From compactness, take $k_0 \in \bigcap_{m=1}^{\infty} \overline{U_m} = \bigcap_{m=1}^{\infty} U_m$. Then $\rho(k_j, k_0) < 1/m$ whenever $j > i_m$, which means that $k_j \to k_0$ in metric ρ. This proves (ii).

(iii) Let Ω be a nonempty open subset of K and fix any $k \in \Omega$. For every $n \in \mathbb{N}$ we find $U_n \in \mathcal{U}_n$ such that $k \in U_n$. Let us assume that $U_n \backslash \Omega \neq \emptyset$ for every $n \in \mathbb{N}$. Since $U_1 \supset U_2 \supset \cdots$, the compactness ensures that the set $\bigcap_{n=1}^{\infty} \overline{U_n} \backslash \Omega$ is nonempty; take some h in it. Then $h \neq k$, and as $\mathcal{U}$ separates the points of K, there is $n \in \mathbb{N}$ such that h does not belong to U_n. However, $U_n \supset \overline{U_{n+1}}$, a contradiction. Hence there is $n \in \mathbb{N}$ so that $U_n \subset \Omega$. Now we realize that U_n coincides with the ρ-open set

$$\{x \in K: \ \rho(k, x) < 1/n\}.$$

This shows that Ω is ρ-open and that the ρ-topology is stronger than the original topology τ.

(iv) Let Ω be the dense G_δ set found in Theorem 5.1.11 for the identity mapping $(K, \tau) \to (K, \tau)$. Then Ω is also a ρ-G_δ set. Now the identity mapping $(\Omega, \tau) \to (\Omega, \rho)$ is continuous, and, by (iii), it is a homeomorphism. The rest follows from (ii) and [En, Theorem 4.3.23].

The sequential compactness of K follows from (iv), Theorem 5.1.10(ii), and Lemma 2.1.1. $\qquad\qquad\square$

If K is a fragmentable compact, then it is in Stegall's class $\mathcal{S}$, and so, by Theorem 3.1.6, K contains a dense G_δ completely metrizable set. It should be noted that (iv) in the above theorem says more.

5.2. CLASS $\widetilde{\mathcal{F}}$ OF BANACH SPACES WITH FRAGMENTABLE DUAL

Definition 5.2.1. The class $\widetilde{\mathcal{F}}$ consists of all Banach spaces V such that (V^*, w^*) belongs to $\mathcal{F}$.

We remark that, according to Theorem 5.1.10(iv), $V \in \widetilde{\mathcal{F}}$ if and only if $(B_{V^*}, w^*) \in \mathcal{F}$.

From Theorems 5.1.11 and 3.2.2 we immediately get the next theorem.

Theorem 5.2.2.

If $V \in \widetilde{\mathcal{F}}$, then $V \in \widetilde{\mathcal{S}}$ and hence V is weak Asplund.

Theorem 5.2.3.

A Banach space V is Asplund if and only if (B_{V^*}, w^*) is fragmented by the metric generated by the dual norm.

Proof. The necessity follows from Theorem 1.1.1. Let us prove the sufficiency. Assume that (B_{V^*}, w^*) is norm fragmented. Let $f : V \to \mathbb{R}$ be a continuous convex function. We have to show that the subdifferential ∂f of f is single-valued and norm-to-norm upper semicontinuous at the points of a residual subset of V; see [Ph, Proposition 2.8]. Take a minimal, norm-to-weak* usco mapping $F : V \to 2^{V^*}$ such that $Fv \subset \partial f(v)$ for all $v \in V$. Consider sets

$$H_n = \left\{ v \in V : \ Fv \cap nB_{V^*} \neq \emptyset \right\}, \qquad n = 1, 2, \ldots;$$

they are closed. Since $\bigcup_{n=1}^\infty H_n = V$, we have $\overline{\bigcup_{n=1}^\infty G_n} = V$, where $G_n = \mathrm{int}H_n$. Take any $n \in \mathbb{N}$. If $G_n = \emptyset$, put $\Omega_n = \emptyset$. Further, assume that $G_n \neq \emptyset$. Since F is minimal usco, Lemma 3.1.2(ii) guarantees that $F(G_n) \subset nB_{V^*}$. Now we know that (nB_{V^*}, w^*) is norm fragmented. Hence, by Proposition 5.1.11, F is single-valued and norm-to-norm upper semicontinuous at each point of a set Ω_n which is residual in G_n. Thus, F is single-valued and norm-to-norm upper semicontinuous at all points of $\bigcup_{n=1}^\infty \Omega_n$. However, this set is residual in V according to Proposition 1.3.2. It then follows, by [Ph, Proposition 2.8.], that f is Fréchet differentiable at each point of this set, which means that V is Asplund. $\square$

Theorem 5.2.4.

(i) If $V \in \widetilde{\mathcal{F}}$ and $T : V \to Y$ is a bounded linear mapping from V into a Banach space Y with dense range, then $Y \in \widetilde{\mathcal{F}}$; in particular, if $V \in \widetilde{\mathcal{F}}$, then every quotient of V belongs to $\widetilde{\mathcal{F}}$.

(ii) If $V \in \widetilde{\mathcal{F}}$, then each subspace of V belongs to $\widetilde{\mathcal{F}}$.

(iii) If $V_n \in \widetilde{\mathcal{F}}$ for every $n = 1, 2, \ldots$, then $\left(\sum_{n=1}^{\infty} V_n \right)_{\ell_1} \in \widetilde{\mathcal{F}}$.

(iv) If V_n, $n = 1, 2, \ldots$, are subspaces of a Banach space V such that $\mathrm{sp}\left(\bigcup_{n=1}^{\infty} V_n \right)$ is dense in V and each $V_n \in \widetilde{\mathcal{F}}$, then $V \in \widetilde{\mathcal{F}}$.

(v) If Γ is a nonempty set and $V_\gamma \in \widetilde{\mathcal{F}}$ for every $\gamma \in \Gamma$, then both $\left(\sum_{\gamma \in \Gamma} V_\gamma \right)_{c_0}$ and $\left(\sum_{\gamma \in \Gamma} V_\gamma \right)_{\ell_p}$, for $p \in (1, +\infty)$, belong to $\widetilde{\mathcal{F}}$.

Proof. (i) Suppose $V \in \widetilde{\mathcal{F}}$, $T : V \to Y$, and $\overline{TV} = Y$. Then $T^* : Y^* \to V^*$ is injective and weak*-to-weak* continuous. Hence, by Theorem 5.1.10(ii), $Y^* \in \widetilde{\mathcal{F}}$.

(ii) Let $V \in \widetilde{\mathcal{F}}$ and let Z be a subspace of V. Then (B_{Z^*}, w^*) is the image of (B_{V^*}, w^*) under the restriction mapping, which is weak*-to-weak* continuous. Now, as $(B_{V^*}, w^*) \in \mathcal{F}$, Theorem 5.1.10(i) applies.

(iii) The assignment $\xi \mapsto \{\xi_{|V_n} : n \in \mathrm{IN}\}$ maps $\left((\sum_{n=1}^{\infty} V_n)_{\ell_1}^*, w^* \right)$ homeomorphically into $\prod_{n=1}^{\infty} (V_n^*, w^*)$, and this last space is fragmentable by Theorem 5.1.10(iii).

(iv) Take V_n as in the premise. Then $\left(\sum_{n=1}^{\infty} V_n \right)_{\ell_1} \in \widetilde{\mathcal{F}}$ by (iii). Now the mapping $\{v_n\} \mapsto \sum_{n=1}^{\infty} v_n$ sends $\left(\sum_{n=1}^{\infty} V_n \right)_{\ell_1}$ linearly and continuously onto a dense set in V. So (i) applies.

(v) Put $Y_0 = \left(\sum_{\gamma \in \Gamma} V_\gamma \right)_{c_0}$ and $Y_p = \left(\sum_{\gamma \in \Gamma} V_\gamma \right)_{\ell_p}$. Similarly, as in (iii), the assignment $\Phi : \xi \mapsto \{\xi_{|V_\gamma} : \gamma \in \Gamma\}$ maps (Y_0^*, w^*) and (Y_p^*, w^*) homeomorphically into $\prod_{\gamma \in \Gamma} (V_\gamma^*, w^*)$. Now, it is well known and easy to prove that Y_0^* is isometric with $\left(\sum_{\gamma \in \Gamma} V_\gamma^* \right)_{\ell_1}$ and Y_p^* is isometric with $\left(\sum_{\gamma \in \Gamma} V_\gamma^* \right)_{\ell_q}$, where $1/p + 1/q = 1$; see [Tay, Example 4.2.4]. Thus, both $\Phi(Y_0^*)$ and $\Phi(Y_p^*)$ lie in the space

$$Z = \left\{ \{v_\gamma^*\} \in \prod_{\gamma \in \Gamma} (V_\gamma^*, w^*) : \{\|v_\gamma^*\|_\gamma\} \in c_0(\Gamma) \right\}.$$

It follows, then, that the proof will be finished once we show that $Z \in \mathcal{F}$. This is done (in a bit more general form) in the next theorem. $\qquad\square$

Theorem 5.2.5.

Let Γ be a nonempty set and assume that for every $\gamma \in \Gamma$ we have a fragmentable space X_γ and a lower semicontinuous function $f_\gamma : X_\gamma \to [0, +\infty)$ such that $f_\gamma^{-1}(0)$ is at most a singleton. Then the space

$$Z = \left\{ \{x_\gamma\} \in \prod_{\gamma \in \Gamma} X_\gamma : \{f_\gamma(x_\gamma)\} \in c_0(\Gamma) \right\}$$

is fragmentable.

Proof. Take any nonempty finite set $F \subset \Gamma$ and denote $X_F = \prod_{\gamma \in F} X_\gamma$. Then $X_F \in \mathcal{F}$ by Theorem 5.1.10(iii). Thus, in X_F, there exist regularly increasing families of open sets $\{R^n_{F,\xi} : \xi \in (0, \xi^n_F]\}$, $n \in \mathbb{N}$, whose union separates the points of X_F; see Proposition 5.1.3, Theorem 5.1.9, and the remark below Definition 5.1.7. By appending $R^n_{F,\xi}$ for some ξ's, we may and do asume that $\xi^n_F = \zeta$ for all $F \subset \Gamma$ and all $n \in \mathbb{N}$. We shall also put $R^n_{F,0} = \emptyset$.

Further, for any nonempty finite set $F \subset \Gamma$ and any $r > 0$ we put

$$W^r_F = \left\{ \{x_\gamma\} \in Z : \{\gamma' \in \Gamma : f_{\gamma'}(x_{\gamma'}) > r\} = F \right\}.$$

Then $W^r_F \cap W^r_H = \emptyset$ whenever $F \neq H$. We also define

$$Q^{r,m} = \left\{ \{x_\gamma\} \in Z : \#\{\gamma' \in \Gamma : f_{\gamma'}(x_{\gamma'}) > r\} > m \right\}, \qquad r > 0, \qquad m \in \mathbb{N};$$

these sets are open in Z, as can easily be checked. We observe that if $F \subset \Gamma$, $\#F = m$, then

$$W^r_F \cup Q^{r,m} = \left\{ \{x_\gamma\} \in Z : f_{\gamma'}(x_{\gamma'}) > r \text{ for every } \gamma' \in F \right\} \cup Q^{r,m}$$

so that this set is open in Z.

For every $m \in \mathbb{N}$ let Γ_m be the family of all sets $F \subset \Gamma$, with $\#F = m$, and let us well order it by a relation $<_m$ [En, Section I.4]. Finally, for $n \in \mathbb{N}$, $m \in \mathbb{N}$, $r > 0$, $F \subset \Gamma_m$, and $\xi \in [0, \zeta]$, we define

$$U^{n,m,r}_{F,\xi} = \bigcup \{W^r_H : H \in \Gamma_m, \ H <_m F\}$$
$$\cup \left[W^r_F \cap \left(R^n_{F,\xi} \times \prod_{\gamma \in \Gamma \setminus F} X_\gamma \right) \right] \cup Q^{r,m};$$

this is an open set in Z. On $\Gamma_m \times [0, \zeta]$, let us consider the lexicographical order. Then for fixed $n \in \mathbb{N}$, $m \in \mathbb{N}$, and $r > 0$, the family

$$\left\{ U^{n,m,r}_{F,\xi} : (F, \xi) \in \Gamma_m \times [0, \zeta] \right\} \cup \{Z\}$$

consists of open sets in Z and is regularly increasing (with respect to the index (F, ξ)). In fact, let us check that (ii) in Definition 5.1.2 is satisfied. Consider any "limit" index (F, α). Then either α is a limit ordinal or $\alpha = 0$ and F is "limit" with respect to the ordering $<_m$. In the first case, (ii) is satisfied since the family $\{R^n_{F,\xi} : \xi \in (0, \zeta]\}$ is regularly increasing. In the second case we have

$$\bigcup \left\{ U^{n,m,r}_{(G,\eta)} : (G, \eta) < (F, 0) \right\} = \bigcup \left\{ U^{n,m,r}_{(G,\zeta)} : G \in \Gamma_m, \ G <_m F \right\}$$
$$= \bigcup \left\{ W^r_G : G \in \Gamma_m, \ G <_m F \right\} \cup Q^{r,m} = U^{n,m,r}_{(F,0)}.$$

Thus, in order to prove that $Z \in \mathcal{F}$, it is enough to show that the family

$$\left\{ U_{F,\xi}^{n,m,r} : \ F \in \Gamma_m, \ \xi \in [0,\zeta], \ n \in \mathbb{N}, \ m \in \mathbb{N}, \ r > 0 \text{ rational} \right\}$$

separates the points of Z. So consider two distinct points $x = \{x_\gamma\}$, $y = \{y_\gamma\}$ in Z.

First assume that $f_\gamma(x_\gamma) = f_\gamma(y_\gamma)$ for all $\gamma \in \Gamma$. We find $\gamma \in \Gamma$ such that $x_\gamma \neq y_\gamma$. Then surely $f_\gamma(x_\gamma) > 0$. We take a rational number r satisfying $f_\gamma(x_\gamma) > r > 0$. Denote $F = \{\gamma' \in \Gamma : f_{\gamma'}(x_{\gamma'}) > r\}$ and $m = \#F$. Then $\gamma \in F$ and so the m-tuples $\{x_{\gamma'} : \gamma' \in F\}$ and $\{y_{\gamma'} : \gamma' \in F\}$ are different. We find $n \in \mathbb{N}$ and $\xi \in (0,\zeta]$ such that the set $R_{F,\xi}^n$ separates them. We observe that $\{x,y\} \cap Q^{r,m} = \emptyset$, $\{x,y\} \subset W_F^r$ and $\{x,y\} \cap W_H^r = \emptyset$ whenever $H \in \Gamma_m$ and $H \neq F$. Therefore the set $U_{F,\xi}^{n,m,r}$ separates the points x and y.

Second, assume there is $\gamma \in \Gamma$ such that $f_\gamma(x_\gamma) \neq f_\gamma(y_\gamma)$. Then there surely exists a rational number $r > 0$ such that

$$F := \{\gamma' \in \Gamma : f_{\gamma'}(x_{\gamma'}) > r\} \neq \{\gamma' \in \Gamma : f_{\gamma'}(y_{\gamma'}) > r\} =: G.$$

We shall consider two cases. First, suppose $\#F = \#G =: m$. Then, say, $F <_m G$. Thus, $x \in W_F^r \subset U_{F,\zeta}^{n,m,r}$. On the other hand, $y \notin W_F^r$ and so $y \notin W_H^r$ for every $H \in \Gamma_m$, with $H <_m F$ and also $y \notin Q^{r,m}$. Hence $y \notin U_{F,\zeta}^{n,m,r}$. Second, suppose $\#F \neq \#G$, say, $\#G < \#F$. Denote $m = \#F$. Then $x \in W_F^r \subset U_{F,\zeta}^{n,m,r}$. Now $y \notin W_H^r$ for every $H \in \Gamma_m$ and $y \notin Q^{r,m}$ as well. Therefore $y \notin U_{F,\zeta}^{n,m,r}$. $\square$

From Theorem 5.2.3 we know that Asplund spaces lie in $\widetilde{\mathcal{F}}$. So Theorem 5.2.4(i), (ii) then yields that subspaces of Asplund generated spaces belong to $\widetilde{\mathcal{F}}$ (and hence to $\widetilde{\mathcal{S}}$). Using symbols we may then write $\mathcal{A} \subset \widetilde{\mathcal{F}} \subset \widetilde{\mathcal{S}}$.

We recall that a compact space K is Eberlein (Radon–Nikodým) compact if and only if $\left(B_{C(K)^*}, w^*\right)$ is; see Theorem 1.2.4 (1.5.4). In what follows, we shall prove the analogous statement for fragmentable compacta.

Lemma 5.2.6.

Let K be a compact space with a separating σ-relatively open partitioning $\mathcal{U} = \bigcup_{n=1}^{\infty} \mathcal{U}_n$ such that $\mathcal{U}_{n+1}$ is a strong refinement of $\mathcal{U}_n$ for every $n = 1, 2, \ldots$. Let H_1 and H_2 be two disjoint nonempty closed subsets of K. Then there exists $U \in \mathcal{U}$ which intersects exactly one of the sets H_1, H_2.

Proof. Write, as usual,

$$\mathcal{U}_n = \{U_\xi^n : \ \xi \in [0, \xi_n)\} \quad \text{and} \quad W_\xi^n = \bigcup_{\eta < \xi} U_\eta^n, \quad \xi \in (0, \xi_n], \quad n \in \mathbb{N}.$$

For every $n \in \mathbb{N}$ and for $i = 1, 2$ put

$$\xi_i^n = \min \left\{ \xi \in [0, \xi_n) : \ U_\xi^n \cap H_i \neq \emptyset \right\} \quad \left(= \min \left\{ \xi \in [0, \xi_n) : \ W_{\xi+1}^n \cap H_i \neq \emptyset \right\} \right).$$

Clearly, we shall be done when we show that $\xi_1^n \neq \xi_2^n$ for some $n \in \mathrm{IN}$. Let us prove this. Since $\mathcal{U}_{n+1}$ is a strong refinement of $\mathcal{U}_n$, we have

$$W_{\xi_i^n}^n \subset W_{\xi_i^{n+1}}^{n+1} \subset W_{\xi_i^{n+1}+1}^{n+1} \subset W_{\xi_i^n+1}^n; \qquad \overline{U_{\xi_i^{n+1}}^{n+1}} \subset U_{\xi_i^n}^n$$

for all $n \in \mathrm{IN}$ and $i = 1, 2$. Hence, by the compactness of K, there exist

$$x_i \in \bigcap_{n=1}^{\infty} U_{\xi_i^n}^n \cap H_i, \qquad i = 1, 2.$$

Note that $x_1 \neq x_2$ as $H_1 \cap H_2 = \emptyset$. Now, we know that $\mathcal{U}$ separates the points of K, so there are $m \in \mathrm{IN}$ and $\xi \in [0, \xi_m)$ such that, say, $x_1 \in U_\xi^m$ and $x_2 \notin U_\xi^m$. Then, surely, $\xi = \xi_1^m$ and $\xi \neq \xi_2^m$. Therefore, $\xi_1^m \neq \xi_2^m$, as desired. $\square$

Given a compact space K, let $M(K)$ denote the Banach space of regular Borel measures on K endowed with the norm

$$\|\mu\| = \sup \left\{ \sum_{i=1}^m |\mu(E_i)| : E_i \subset K \text{ are mutually disjoint Borel sets} \right\}, \quad \mu \in M(K).$$

We recall that, according to F. Riesz's representation theorem, $M(K)$ and $C(K)^*$ are linearly isometric; in what follows we shall identify these two spaces.

Theorem 5.2.7.

A compact K is fragmentable (if and) only if $\left(B_{C(K)^*}, w^*\right)$ is fragmentable, that is, $C(K) \in \tilde{\mathcal{F}}$.

Proof. The sufficiency is a consequence of Theorem 5.1.10(ii), so let us focus on the necessity. Let K be a compact fragmentable space. Denote by $\mathcal{M}^+$ the set of all $\mu \in M(K)$ such that $\mu(E) \geq 0$ for every Borel set $E \subset K$. We observe that once $(\mathcal{M}^+, w^*)$ is fragmentable, then so is the whole of $(M(K), w^*)$. In fact, repeatedly using Proposition 5.1.10, we have that $\mathcal{M}^+ \cap B_{M(K)} \in \mathcal{F}$, $(\mathcal{M}^+ \cap B_{M(K)}) \times (\mathcal{M}^+ \cap B_{M(K)}) \in \mathcal{F}$, $(\mathcal{M}^+ \cap B_{M(K)}) - (\mathcal{M}^+ \cap B_{M(K)}) \in \mathcal{F}$, and $B_{M(K)} \in \mathcal{F}$ by [DS, Corollary III.4.11], and finally $M(K) = \bigcup_{n=1}^{\infty} n B_{M(K)} \in \mathcal{F}$. Therefore $\left(B_{C(K)^*}, w^*\right) \in \mathcal{F}$.

Now let $\mathcal{U} = \bigcup_{n=1}^{\infty} \mathcal{U}_n$ be a separating σ-relatively open partitioning of K such that $\mathcal{U}_{n+1}$ is a strong refinement of $\mathcal{U}_n$ for each $n \in \mathrm{IN}$; see Theorem 5.1.9 and Proposition 5.1.8. Write

$$\mathcal{U}_n = \left\{ U_\xi^n : \xi \in [0, \xi_n) \right\} \qquad \text{and} \qquad W = \left\{ W_\xi^n : \xi \in (0, \xi_n], \ n \in \mathrm{IN} \right\},$$

where $W_\xi^n = \bigcup_{\eta < \xi} U_\eta^n$. It is natural to define the following families of sets in $\mathcal{M}^+$:

$$\mathcal{W}_{nr} = \left\{ W_\xi^{nr} : \xi \in (0, \xi_n + 1] \right\}, \qquad n = 1, 2, \ldots, \qquad r > 0,$$

where $W^{nr}_{\xi_n+1} = \mathcal{M}^+$ and

$$W^{nr}_\xi = \{\mu \in \mathcal{M}^+ : \mu(W^n_\xi) > r\}, \qquad \xi \in (0, \xi_n].$$

The sets W^{nr}_ξ are weak* open in $\mathcal{M}^+$. Indeed, let $\mu \in W^{nr}_\xi$. From the regularity of μ we can find a compact set $C \subset W^n_\xi$ such that $\mu(C) > r$. Further, there is a function $f \in C(K)$ such that $0 \le f \le 1$, $f(k) = 1$ for $k \in C$ and $f(k) = 0$ for $k \in K \backslash W^n_\xi$ [En, Theorem 3.1.7]. Then $\langle \mu, f \rangle > r$ and the weak* open set $\{\nu \in \mathcal{M}^+ : \langle \nu, f \rangle > r\}$ contains μ and is contained in W^{nr}_ξ. This shows that W^{nr}_ξ is an open set. We note that each $\mathcal{W}_{nr}$ is a regularly increasing family of open sets in $(\mathcal{M}^+, w^*)$. Indeed, take any limit ordinal $\xi \in (0, \xi_n]$ and $\mu \in W^{nr}_\xi$. From the regularity of μ, we find a compact set $C \subset W^n_\xi$ such that $\mu(C) > r$. Since $\bigcup_{\eta < \xi} W^n_\eta = W^n_\xi$, there is $\eta < \xi$ such that $C \subset W^n_\eta$ and so $\mu(W^n_\eta) > r$. Thus $\mu \in W^{nr}_\eta$.

According to the remark below Definition 5.1.7, it remains to prove that the family $\bigcup \{\mathcal{W}_{nr} : n \in \mathbb{N}, r > 0 \text{ rational}\}$ separates the points of $\mathcal{M}^+$. So let us assume that two distinct $\mu, \nu \in \mathcal{M}^+$ are given. It is sufficient to find $n \in \mathbb{N}$ and $\xi \in (0, \xi_n]$ such that $\mu(W^n_\xi) \ne \nu(W^n_\xi)$. Indeed, then, for some rational $r > 0$, we have that $\{\mu, \nu\} \cap W^{nr}_\xi$ is a singleton. So let us assume, by contradiction, that $\mu(W) = \nu(W)$ for all $W \in \mathcal{W}$. Since $\mu \ne \nu$, there is an open set $U \subset K$ so that, say, $\mu(U) > \nu(U)$. We may then find a compact set $K_1 \subset U$ such that $\mu(K_1) > \nu(U)$. Put $K_2 = K \backslash U$; then $K_1 \cap K_2 = \emptyset$. In what follows we shall find a set $D \subset K$ which is both μ and ν measurable and is such that $D \supset K_1$, $D \cap K_2 = \emptyset$, and $\mu(D) = \nu(D)$. Having such a set we shall obtain

$$\nu(D) = \mu(D) \ge \mu(K_1) > \nu(U) \ge \nu(D),$$

which is impossible. This will finish the proof.

In order to find the set D, we shall construct an increasing family $\{G_\xi : 0 < \xi \le \xi_0\}$ of open sets in K such that each G_ξ is the union of elements from $\mathcal{W}$. Put $G_1 = \emptyset$. Consider an ordinal $\xi > 1$ and assume we have already found G_η for every $\eta < \xi$. If ξ is a limit ordinal, put $G_\xi = \bigcup_{\eta < \xi} G_\eta$. Next, let ξ have a predecessor, say, $\xi = \eta + 1$. If G_η contains K_1 or K_2, put $G_\xi = K$ and stop the process. Otherwise, by applying Lemma 5.2.6 for $H_i = K_i \backslash G_\eta$, $i = 1, 2$, we find $G_\xi \in \mathcal{W}$ such that G_ξ intersects exactly one of the sets H_1, H_2. Since $\mathcal{U}_{n+1}$ is a refinement of $\mathcal{U}_n$ for every $n \in \mathbb{N}$, we can easily deduce that $G_\xi \supset G_\eta$, and as $G_\xi \backslash G_\eta$ is a nonempty set, our process must eventually stop. Note that the regularity of μ and ν and the coincidence of them on the family $\mathcal{W}$ ensures that $\mu(G_\xi) = \nu(G_\xi)$ for all $\xi \in (0, \xi_0]$.

Denote

$$\Delta = \{\xi \in (0, \xi_0) : (G_{\xi+1} \backslash G_\xi) \cap K_1 \ne \emptyset\}.$$

Thus the set $D := \bigcup_{\xi \in \Delta} (G_{\xi+1} \backslash G_\xi)$ contains K_1 and does not intersect K_2. It remains to show that D is both μ and ν measurable and that $\mu(D) = \nu(D)$. To do this, put

$$D_\xi = D \cap G_\xi, \qquad \xi \in (0, \xi_0].$$

By transfinite induction, we shall show that D_ξ is μ and ν measurable and that $\mu(D_\xi) = \nu(D_\xi)$ for all $\xi \in (0, \xi_0]$. Thus we shall, in particular, get $\mu(D) = \nu(D)$ and the proof will be finished. Trivially, $D_1 (= \emptyset)$ is a Borel set and $\mu(D_1) = \nu(D_1)$. Consider $\zeta > 1$ and assume that the induction claim has already been verified for every $\xi < \zeta$. If ζ is a nonlimit ordinal, then we can immediately verify the induction claim for $\xi = \zeta$. Further, let ζ be a limit ordinal. Denote $\lambda = \mu + \nu$. Trivially, λ is a regular Borel measure and D_ξ, $\xi < \zeta$, are λ measurable. From the regularity of λ, we can easily deduce that $\lambda(G_\zeta) = \lim_{\xi \uparrow \zeta} \lambda(G_\xi)$. Moreover, we can find $\xi_1 < \xi_2 < \cdots < \zeta$ so that $\lambda(G_\zeta) = \lim_{n \to \infty} \lambda(G_{\xi_n})$. Thus, we have

$$D_{\xi_n} \subset D_\zeta \subset \left(G_\zeta \backslash G_{\xi_n} \right) \cup D_{\xi_n}$$

and

$$\lim_{n \to \infty} \lambda \left(D_{\xi_n} \right) \leq \lim_{n \to \infty} \lambda \left[\left(G_\zeta \backslash G_{\xi_n} \right) \cup D_{\xi_n} \right] = \lim_{n \to \infty} \lambda \left(D_{\xi_n} \right).$$

Hence, putting

$$D_0 = \bigcup_{n=1}^{\infty} D_{\xi_n} \quad \text{and} \quad E_0 = \bigcap_{n=1}^{\infty} \left[\left(G_\zeta \backslash G_{\xi_n} \right) \cup D_{\xi_n} \right],$$

these sets are λ measurable, $D_0 \subset D_\zeta \subset E_0$, and $\lambda(D_0) = \lambda(E_0)$. It then follows that D_ζ is also λ measurable and $\lambda(D_\zeta) = \lim_{n \to \infty} \lambda(D_{\xi_n})$. From this we can easily deduce that D_ζ is both μ and ν measurable and that

$$\mu(D_\zeta) = \lim_{n \to \infty} \mu(D_{\xi_n}) \quad \text{and} \quad \nu(D_\zeta) = \lim_{n \to \infty} \nu(D_{\xi_n}).$$

Therefore

$$\mu(D_\zeta) = \lim_{n \to \infty} \mu(D_{\xi_n}) = \lim_{n \to \infty} \nu(D_{\xi_n}) = \nu(D_\zeta).$$

$$\square$$

5.3. *GÂTEAUX SMOOTH SPACES BELONG TO* $\widetilde{\mathcal{F}}$

Theorem 5.3.1.

Let a Banach space V have a Gâteaux smooth norm $\| \cdot \|$. Then $V \in \widetilde{\mathcal{F}}$.

Proof. Throughout the argument we shall use the same symbol for a norm on V and its dual norm on V^*. It is enough to show that $(B_{V^*}, w^*) \in \mathcal{F}$. Consider a nonempty subset M of B_{V^*}, $\epsilon > 0$, $\beta > 0$, and a fixed equivalent

Gâteaux smooth norm p on V. We shall construct a relatively open partitioning $\mathcal{U} = \{U_\zeta : \zeta \in [0, \xi)\}$ of (B_{V^*}, w^*), $s_\zeta \geq 0$, $e_\zeta \in V$, and equivalent Gâteaux smooth norms p_ζ on V, $\zeta \in [0, \xi)$, as follows. Put $U_0 = \emptyset$, $s_0 = 0$, $p_0 = p$ and let e_0 be any element of V with $p(e_0) = 1$. Let $\zeta > 0$ be some ordinal and assume we have already found U_η, s_η, e_η, and p_η for all $0 \leq \eta < \zeta$. Put $R = M \setminus \bigcup_{\eta < \zeta} U_\eta$. If $R = \emptyset$, put $\xi = \zeta$ and stop the process. Otherwise put $s_\zeta = \sup\{p(v^*) : v^* \in R\}$. If $s_\emptyset = 0$, put $U_\zeta = \{0\}$ and let e_ζ be any element of V with $p(e_\zeta) = 1$. Further, assume $s_\zeta > 0$. We find $e_\zeta \in V$ such that $p(e_\zeta) = 1$ and the set $\{v^* \in R : \langle v^*, e_\zeta \rangle > (1 - \epsilon)s_\zeta\}$ is nonempty; denote it by U_ζ. Then U_ζ is a weak* relatively open set in R and

$$\sup\{p(v^*) : v^* \in B_{V^*}\} \geq s_\zeta \geq \sup\{p(v^*) : v^* \in U_\zeta\}$$
$$\geq \inf\{\langle v^*, e_\zeta \rangle : v^* \in U_\zeta\} \geq (1 - \epsilon)s_\zeta.$$

Finally define p_ζ by

$$p_\zeta^2 = p^2 + \beta^2 \mathrm{dist}_{\|\cdot\|}(\cdot, \mathbb{R}e_\zeta)^2,$$

where $\mathrm{dist}_{\|\cdot\|}(\cdot, \mathbb{R}e_\zeta)$ means the $\|\cdot\|$-distance from the line $\mathbb{R}e_\zeta$. Obviously, p_ζ is an equivalent norm on V and $p_\zeta(e_\zeta) = p(e_\zeta) = 1$. This finishes the induction step. Clearly, after finishing the process, we get that $\mathcal{U} := \{U_\zeta : \zeta \in [0, \xi)\}$ is a relatively open partitioning of (B_{V^*}, w^*). In what follows it will be useful to use a new notation: If $U \in \mathcal{U}$, we put $s_U = s_\zeta$, $e_U = e_\zeta$, and $p_U = p_\zeta$, where $\zeta \in [0, \xi)$ is such that $U = U_\zeta$. Let us summarize the properties of such s_U, e_U, and p_U: For each $U \in \mathcal{U}$ we have

$$\sup\{p(v^*) : v^* \in B_{V^*}\} \geq s_U \geq \sup\{p(v^*) : v^* \in U\}$$
$$\geq \inf\{\langle v^*, e_U \rangle : v^* \in U\} \geq (1 - \epsilon)s_U, \tag{1}$$

$$p_U(e_U) = p(e_U) = 1, \tag{2}$$

$$p_U^2 = p^2 + \beta^2 \mathrm{dist}_{\|\cdot\|}(\cdot, \mathbb{R}e_U)^2. \tag{3}$$

Now we shall apply the above construction countably many times in order to produce relatively open partitionings $\mathcal{U}_1$, $\mathcal{U}_2$, ... of (B_{V^*}, w^*) such that their union will be separating. Thus we shall prove, according to Theorem 5.1.9, that (B_{V^*}, w^*) is fragmentable and hence $V \in \tilde{\mathcal{F}}$. Choose sequences $\{\epsilon_n\}$, $\{\beta_n\}$ of positive numbers such that

$$\tfrac{1}{2} > \epsilon_1 > \epsilon_2 > \cdots, \quad \sum_{n=1}^{\infty} \beta_n^2 < 3, \quad \text{and} \quad \sum_{n=1}^{\infty} \frac{\sqrt{\epsilon_n}}{\beta_n} < +\infty;$$

then $\epsilon_n \to 0$ as $n \to \infty$. Let $\mathcal{U}_1$ be the family $\mathcal{U}$ from the first paragraph obtained for $M = B_{V^*}$, $\epsilon = \epsilon_1$, $\beta = \beta_1$, and $p = \|\cdot\|$. Of course, we also get, this way, s_U, e_U, and p_U for each $U \in \mathcal{U}_1$. Now fix any $U \in \mathcal{U}_1$ and apply the first paragraph to $M = U$, $\epsilon = \epsilon_2$, $\beta = \beta_2$, and $p = p_U$. We get a relatively open partitioning $\mathcal{U}^U$ of (U, w^*) and s_W, e_W, p_W for every $W \in \mathcal{U}^U$. Performing this for every $U \in \mathcal{U}_1$ we put $\mathcal{U}_2 = \bigcup \{\mathcal{U}^U : U \in \mathcal{U}_1\}$ and endow it with the lexicographical order. According to Proposition 5.1.5, $\mathcal{U}_2$ is a relatively open partitioning of (B_{V^*}, w^*). Now repeat the same process for every $U \in \mathcal{U}_2$ and for $\epsilon = \epsilon_3$, $\beta = \beta_3$. Then we get a relatively open partitioning $\mathcal{U}_3$ of (B_{V^*}, w^*) and s_W, e_W, p_W for every $W \in \mathcal{U}_3$. It should now be clear how to proceed in general. In this way we get, for each $n \in \mathbb{N}$, a relatively open partitioning $\mathcal{U}_n$ of (B_{V^*}, w^*) and corresponding s_W, e_W, p_W for each $W \in \mathcal{U}_n$.

We claim the family $\mathcal{U} := \bigcup_{n=1}^{\infty} \mathcal{U}_n$ is separating. By contradiction, assume this is not true. Then there are v_1^*, $v_2^* \in B_{V^*}$, $v_1^* \neq v_2^*$, which are not separated by $\mathcal{U}$. Hence for each $n = 1, 2, \ldots$ there is $U_n \in \mathcal{U}_n$ such that $v_1^*, v_2^* \in U_n$. Clearly $U_1 \supset U_2 \supset \cdots \supset \{v_1^*, v_2^*\}$. Denote, for brevity, $s_n = s_{U_n}$, $e_n = e_{U_n}$, $\|\cdot\|_n = p_{U_n}$. From (1), (2), and (3) we get

$$\sup \{\|v^*\|_{n-1} : \ v^* \in B_{V^*}\} \geq s_n \geq \sup \{\|v^*\|_{n-1} : \ v^* \in U_n\}$$
$$\geq \inf \{\langle v^*, e_n \rangle : \ v^* \in U_n\} \geq (1 - \epsilon_n) s_n, \qquad (4)$$
$$\|e_n\|_n = \|e_n\|_{n-1} = 1,$$
$$\|\cdot\|_n^2 = \|\cdot\|_{n-1}^2 + \beta_n^2 \mathrm{dist}_{\|\cdot\|}(\cdot, \mathbb{R}e_n)^2$$

for all $n = 2, 3, \ldots$. The rest of this proof is almost identical with that of Theorem 4.2.4. Remarking that

$$\|\cdot\|^2 \leq \|\cdot\|_n^2 \leq \|\cdot\|_{n+1}^2 \leq \left(1 + \sum_{j=1}^{n+1} \beta_j^2\right) \|\cdot\|^2,$$

we get that $\|\cdot\|_n$ converge uniformly on bounded sets to some $\|\cdot\|_\infty$. Then $\|\cdot\|_\infty$ is Gâteaux smooth. Further, the sequence $\{e_n\}$ can be shown to converge to some $e_\infty \in V$ with $\|e_\infty\|_\infty = 1$. Finally, (4) ensures that

$$\langle v_1^*, e_\infty \rangle = \lim_{n \to \infty} s_n = \langle v_2^*, e_\infty \rangle.$$

Hence, the Gâteaux smoothness of $\|\cdot\|_\infty$ yields that $v_1^* = v_2^*$, a contradiction. So it follows that the family $\mathcal{U}$ separates the points of B_{V^*}. $\qquad\square$

5.4. NOTES AND REMARKS

Fragmentability was introduced by Jayne and Rogers [JR]. In Sections 5.1 and 5.2, we followed the paper of Ribarska [Ri1]. We refer to Stegall's paper [St12] for many other results related to Chapters 3 and 5. A concept very close to our regularly increasing family is called there a *woof*. Theorems 5.2.4(v) and 5.2.5 are from [St14]. We do not know if $\widetilde{S}$ in Theorem 3.2.3, part (ii) or (vi), can be replaced by $\widetilde{\mathcal{F}}$. That belonging of a space to the class $\widetilde{\mathcal{F}}$ is not a three-space property follows from Theorem 2.3.1.

Theorem 5.3.1 is again due to Ribarska [Ri2]. For an easier understanding of her proof the reader is recommended to consider first a norm whose dual norm is strictly convex; see the proof of Proposition 4.2.2. Then the gymnastics with subsequent improvements of the norm can be avoided, and we simply deduce that $\|v_1^*\| = \|v_2^*\| = \frac{1}{2}\|v_1^* + v_2^*\|$. Hence the strict convexity ensures that $v_1^* = v_2^*$. A prehistory of this theorem is as follows. The technique of subsequent renormings is due to Preiss [Pr]. It was then used in [PPN], see Section 4.2, and from this the proof of Theorem 5.3.1 stems.

There is a simple way to prove that the unit ball in a strictly convex Banach space with the weak topology is fragmentable. This was shown to us by W. B. Moors and S. Sciffer. Let $(V, \|\cdot\|)$ be such a space. We define $\tau : B_V \times B_V \to [0, +\infty)$ by

$$\tau(u, v) = \max(\|u\|, \|v\|) - \tfrac{1}{2}\|u + v\|, \qquad u, v \in B_V.$$

(This is not a metric.) The strict convexity of $\|\cdot\|$ immediately guarantees that $\tau(u, v) = 0$ implies $u = v$. Consider any $\emptyset \neq M \subset B_V$ and any $\epsilon > 0$. Denote $\alpha = \sup\{\|v\| : v \in M\}$. Then there is $\xi \in B_{V^*}$ such that the slice $S := \{v \in M : \langle \xi, v \rangle > \alpha - \epsilon\}$ is nonempty. Observe also that S is relatively weakly open in B_V. Take now any $u, v \in S$. Then $\tau(u, v) \leq \max(\|u\|, \|v\|) - \langle \xi, \frac{1}{2}(u + v) \rangle < \alpha - (\alpha - \epsilon) = \epsilon$. Therefore τ fragments (B_V, w). Now, by the remark after the proof of Theorem 5.1.9, we can conclude that (B_V, w) is fragmentable. An analogous reasoning yields that $(B_{V^*}, w^*) \in \mathcal{F}$ provided that the dual norm on V^* is strictly convex.

A certain raison d'être for Theorem 5.3.1 are Talagrand's results: *The (Asplund) space $C([0, \omega_1])$ admits an equivalent Fréchet smooth norm (even a C^∞ smooth norm by Haydon [Ha6]) while its dual does not even admit an equivalent strictly convex dual norm* [Ta5; DGZ2, Section VII.5].

There exists an extension of Theorem 5.3.1 which was stated in [DGZ1] and proved by Fosgerau in his thesis [Fos]: $V \in \widetilde{\mathcal{F}}$ *if there exists a Lipschitz function* $\varphi : V \to \mathbb{R}$, *with bounded nonempty support, which is everywhere Gâteaux differentiable or just with the property that* $\lim_{t \downarrow 0}[\varphi(v + th) + \varphi(v - th) - 2\varphi(v)]/t \geq 0$ *for every $v \in V$ and every $h \in V$.* The utility of this extension is supported by Haydon's counterexample [Ha1, Ha6; DGZ2, Sections VI.9, VII.7]. He took a full uncountably branching tree T of height ω_1 and showed that *the (Asplund) space $C_0(T)$ of continuous functions on T, vanishing "at infinity," does not admit an equivalent Gâteaux smooth norm and yet possesses a C^∞ smooth function with bounded nonempty support.* Hence *fragmentability of (V^*, w^*) does not always imply Gâteaux smoothness of V.*

Theorems 5.3.1, 4.2.3, and 4.2.4 raise another natural question: If V^* has an equivalent, not necessarily dual, strictly convex norm, is V then in $\widetilde{\mathcal{F}}$, in $\widetilde{\mathcal{S}}$, or just weak Asplund? This was answered negatively by Argyros and Mercourakis [AM]. They constructed a *WLD space V* (see Definition 7.2.6), *with an unconditional basis, which is not weak Asplund and yet V^* has an equivalent strictly convex (nondual) norm.* Thus (B_{V^*}, w^*) is a Corson compact (see

Definition 7.2.6) not belonging to $\mathcal{F}$. Probably it is unknown *if weak Asplundness implies that the dual admits an equivalent strictly convex (nondual) norm.* Here we only mention a strong analogue of this: *If V is Asplund, then V^* admits an equivalent locally uniformly rotund (nondual) norm* [FG].

In a quite recent paper of Kenderov and Moors [KM], it is shown that fragmentability has another topological characterization: It is equivalent to the existence of a winning strategy for player **B** in a Banach–Mazur-like game.

During reading the copyedited manuscript of this book, O. Kalenda informed us that he had constructed, under some set-theoretical assumptions, an example of *a nonfragmentable compact space K belonging to the Stegall's class $\mathcal{S}$.* He does not know if $C(K) \in \widetilde{\mathcal{S}}$ for this K. The question *whether the class $\widetilde{\mathcal{F}}$ is strictly smaller than the class $\widetilde{\mathcal{S}}$ is still open.*

Chapter Six

"Long Sequences" of Linear Projections

This chapter has an auxiliary character. It is devoted to the construction of "long sequences" of linear, norm 1 projections with some nice properties. Such a sequence is said to be a projectional resolution of the identity (P.R.I.) and enables us to split the space into many "smaller" subspaces. In this way, we obtain a good technique for proving various properties of Banach spaces via a transfinite induction argument. A starting point for the construction of a P.R.I. will be the so-called projectional generator. Yet, its existence will not be investigated here. This will be done in the next two chapters. In fact, we shall show there that in two quite different classes of Banach spaces the projectional generators do exist, and thus, we shall obtain some interesting deeper results.

The first section is devoted to the construction of a P.R.I. from the projectional generator. In the second section, we show that the existence of a P.R.I. on a given Banach space is useful for the construction of an injection into $c_0(\Gamma)$, of a (shrinking) Markuševič basis, and in proving that a given Banach space is WCG. A "separable" variant of the P.R.I. is also considered here.

6.1. A GENERAL METHOD FOR CONSTRUCTING A P.R.I.

Let V be a Banach space. For $A \subset V$, $B \subset V^*$ we denote

$$A^{\perp} = \{v^* \in V^* : \langle v^*, a \rangle = 0 \ \text{ for all } \ a \in A\},$$

$$B_{\perp} = \{v \in V : \langle b, v \rangle = 0 \ \text{ for all } \ b \in B\}.$$

The following equivalence of the concept of projection will be useful.

Lemma 6.1.1.

Let $P : V \to V$ be a linear norm 1 projection (i.e., P is linear, $\|P\| = 1$, and $P \circ P = P$) on a Banach space V and put $E = PV$ and $F = P^* V^*$. Then
  (i) $\|v\| = \sup\langle F \cap B_{V^*}, v \rangle$ for all $v \in E$ and
  (ii) $E^\perp \cap F = \{0\}$.
  Conversely, suppose there exist two sets $A \subset V$, $B \subset V^*$ such that $\overline{A}$, $\overline{B}$ are linear and
  (i') $\|a\| = \sup\langle B \cap B_{V^*}, a \rangle$ for all $a \in A$ and
  (ii') $A^\perp \cap \overline{B}^* = \{0\}$,
Then there exists a linear norm 1 projection $P : V \to V$ such that $PV = \overline{A}$, $P^{-1}(0) = B_\perp$ and $P^* V^* = \overline{B}^*$.

Proof. If $v \in PV$, then

$$\|v\| = \|Pv\| = \sup\langle B_{V^*}, Pv \rangle = \sup\langle P^* V^* \cap B_{V^*}, v \rangle = \sup\langle F \cap B_{V^*}, v \rangle.$$

If $\xi \in E^\perp \cap F$, then for every $v \in V$

$$\langle \xi, v \rangle = \langle P^* \xi, v \rangle = \langle \xi, Pv \rangle = 0$$

and so $\xi = 0$.
  Now let A, B be as in the second part of the lemma. If $v \in \overline{A} \cap B_\perp$, then, by (i'),

$$\|v\| = \sup\langle B \cap B_{V^*}, v \rangle = 0.$$

Moreover, $\overline{A} + B_\perp$ is closed since we have, for all $a \in A$ and all $u \in B_\perp$,

$$\|a + u\| \geq \sup\langle B \cap B_{V^*}, a + u \rangle = \sup\langle B \cap B_{V^*}, a \rangle = \|a\|$$

according to (i'). Finally, if $\overline{A} + B_\perp \neq V$, then by the separation theorem, there is $0 \neq \xi \in V^*$ which is identically zero on $\overline{A} + B_\perp$. Therefore $\xi \in A^\perp \cap \overline{B}^*$ by the separation theorem and [DS, Theorem V.3.9], and so, by (ii'), $\xi = 0$, a contradiction. We have thus shown that V is a direct sum of $\overline{A}$ and $B_\perp$.
  Define $P : V \to V$ by

$$P(a + u) = a, \qquad a \in \overline{A}, \qquad u \in B_\perp.$$

Then P is a linear projection with $\|P\| = 1$, $PV = \overline{A}$, and $P^{-1}(0) = B_\perp$. If $\xi \in B$, then for all $v \in V$ we have $\langle P^* \xi, v \rangle = \langle \xi, Pv \rangle = \langle \xi, v \rangle$ as $Pv - v \in B_\perp$; so $\xi \in P^* V^*$. Hence, $\overline{B}^* \subset P^* V^*$. Assume that there is $\xi \in P^* V^* \backslash \overline{B}^*$. Using the separation theorem and [DS, Theorem V.3.9], we find $v \in V$ such that $\langle \xi, v \rangle \neq 0 = \sup\langle B, v \rangle$. (Here, $\bar{B}$ is linear.) It follows, then, that $v \in B_\perp$ and so $Pv = 0$. However, $0 \neq \langle \xi, v \rangle = \langle \xi, Pv \rangle$, a contradiction. $\qquad\square$

Sometimes it is important to know that the condition (ii') can be replaced by a weaker one:

Lemma 6.1.2.

Let $A \subset V$, $B \subset V^*$ *be such that* $\bar{A}$, $\bar{B}$ *are linear and*
(i') $\|a\| = \sup\langle B \cap B_{V^*}, a \rangle$ *for all* $a \in A$ *and*
(ii'') $A^\perp \cap \overline{B \cap B_{V^*}}^* = \{0\}$.
Then
(ii') $A^\perp \cap \bar{B}^* = \{0\}$.

Proof. Put $Y = \bigcup_{n=1}^\infty n\overline{B \cap B_{V^*}}^*$. Since $\bar{B}$ is linear, Y is also linear. We shall show that Y is weak* closed and hence $Y = \bar{B}^*$. From this (ii') will immediately follow. Denote $E = \bar{A}$ and let $Q : V^* \to E^*$ be the restriction mapping, that is,

$$\langle Qv^*, e \rangle = \langle v^*, e \rangle, \qquad v^* \in V^*, \qquad e \in E.$$

Then Q is injective on Y. For if $\xi \in Y$ and $Q\xi = 0$, we have $\xi \in A^\perp$. Moreover, $(1/n)\xi \in \overline{B \cap B_{V^*}}^*$ for some $n \in \text{IN}$. Hence, by (ii''), $\xi = 0$. Further, we shall check that $Q(\overline{B \cap B_{V^*}}^*) = B_{E^*}$. Obviously, the inclusion "$\subset$" holds. Assume that the reverse inclusion is not true. Then, since $Q(\overline{B \cap B_{V^*}}^*)$ is a weak* closed convex set, the separation theorem and [DS, Theorem V.3.9] provide us with $e \in E$ and $e^* \in B_{E^*}$ such that $\sup\langle Q(\overline{B \cap B_{V^*}}^*), e \rangle < \langle e^*, e \rangle$. Hence, by (i'), $\|e\| < \langle e^*, e \rangle$, a contradiction. Finally, we claim that $Y \cap B_{V^*} = \overline{B \cap B_{V^*}}^*$. Having proved this, another Krein–Šmulyan theorem [DS, Theorem V.5.7] will then guarantee that Y is weak* closed. In order to prove the claim, we check that

$$Q(Y \cap B_{V^*}) \subset Q(B_{V^*}) = B_{E^*} = Q(\overline{B \cap B_{V^*}}^*).$$

Since Q is injective on Y, we get that $Y \cap B_{V^*} \subset \overline{B \cap B_{V^*}}^*$. The reverse inclusion is trivial. $\qquad\square$

The next two statements will provide us with a very basic and universal tool for constructing "long sequences" of linear projections with nice properties.

Lemma 6.1.3.

Let V be a Banach space and $Y \subset V$ its subspace. Let W be a subset of V^* with $\overline{W}$ linear and let $\Phi : W \to 2^V$ and $\Psi : V \to 2^W$ be two, at most countable valued mappings. Finally let $\aleph$ be an infinite cardinal number and let $A_0 \subset V$, $B_0 \subset W$ be two subsets with $\#A_0 \leq \aleph$, $\#B_0 \leq \aleph$. Then there exist sets $A_0 \subset A \subset V$, $B_0 \subset B \subset W$ such that $\overline{A} \cap Y = \overline{A \cap Y}$, $\overline{A}$, $\overline{B}$ are linear, $\#A \leq \aleph$, $\#B \leq \aleph$, and $\Phi(B) \subset A$, $\Psi(A) \subset B$.

Proof. Let $f : V \to Y$ be a mapping assigning to each $v \in V$ a point $f(v) \in Y$ such that $\|v - f(v)\| \leq 2 \operatorname{dist}(v, Y)$. We shall use an old glueing argument due to Mazur. By induction, we shall construct sequences of sets $A_0 \subset A_1 \subset A_2 \subset \cdots \subset V$ and $B_0 \subset B_1 \subset B_2 \subset \cdots \subset W$ as follows. Assume that for some fixed $n \in \mathrm{IN}$ we have found A_i, B_i, $i = 0, 1, \ldots, n-1$. Then put

$$A_n = \left\{ \sum_{i=1}^m r_i v_i : \ v_i \in A_{n-1} \cup f(A_{n-1}) \cup \Phi(B_{n-1}), \right.$$

$$\left. r_i \text{ are rational}, \ i = 1, \ldots, m, \ m \in \mathrm{IN} \right\}$$

and

$$B_n = \left\{ \sum_{i=1}^m r_i v_i^* : \ v_i^* \in B_{n-1} \cup \Psi(A_{n-1}), \ r_i \text{ are rational}, \ i = 1, \ldots, m, \ m \in \mathrm{IN} \right\}.$$

Now put $A = \bigcup_{n=1}^\infty A_n$, $B = \bigcup_{n=1}^\infty B_n$. If $a_1, a_2 \in A$, then $a_1, a_2 \in A_n$ for some $n \in \mathrm{IN}$ and so $a_1 + a_2 \in A_{n+1} \subset A$. Similarly, if $a \in A$ and $\lambda \in \mathrm{IR}$, there is $n \in \mathrm{IN}$ so that $\lambda a \in \lambda A_n \subset \overline{A_{n+1}} \subset \overline{A}$, which shows that $\overline{A}$ is linear. An analogous argument guarantees the linearity of $\overline{B}$. We shall show that $\overline{A} \cap Y = \overline{A \cap Y}$. Let $y \in \overline{A} \cap Y$; so there is a sequence $\{a_i\} \subset A$ converging to y. For each $i \in \mathrm{IN}$ we find $n_i \in \mathrm{IN}$ so that $a_i \in A_{n_i}$. Then

$$\|f(a_i) - y\| \leq \|f(a_i) - a_i\| + \|a_i - y\|$$
$$\leq 2 \operatorname{dist}(a_i, Y) + \|a_i - y\| \leq 3\|a_i - y\| \to 0$$

as $i \to \infty$. Since $f(a_i) \in f(A_{n_i}) \subset A_{n_i+1} \subset A$, it follows that $y \in \overline{A \cap Y}$ and so $\overline{A} \cap Y \subset \overline{A \cap Y}$. The reverse inclusion always holds. The remaining properties of the sets A and B claimed in the statement of the lemma are almost obvious. $\qquad\square$

The above lemma can be used repeatedly in a transfinite induction argument to obtain the following:

Proposition 6.1.4.

Let V, Y, W, Φ, and Ψ be as in Lemma 6.1.3. Assume that dens $Y > \aleph_0$ and let μ be the first ordinal with card $\mu = $ dens Y. Then there exist "long sequences", that is, families $\{A_\alpha : \omega < \alpha \le \mu\}$ and $\{B_\alpha : \omega < \alpha \le \mu\}$ of subsets in V and W, respectively, such that $\overline{A_\mu} \supset Y$ and for each $\omega < \alpha \le \mu$ the following holds:

$\overline{A_\alpha}$, $\overline{B_\alpha}$ are linear,

$\overline{A_\alpha} \cap Y = \overline{A_\alpha \cap Y}$,

$\#A_\alpha \le $ card α, $\#B_\alpha \le $ card α,

$\Phi(B_\alpha) \subset A_\alpha$, $\Psi(A_\alpha) \subset B_\alpha$,

$A_\beta \subset A_\alpha$, $B_\beta \subset B_\alpha$ if $\omega < \beta \le \alpha$, and

$A_\alpha = \bigcup_{\beta < \alpha} A_{\beta+1}$, $B_\alpha = \bigcup_{\beta < \alpha} B_{\beta+1}$.

Proof. Let f be defined as in the proof of Lemma 6.1.3. Let $\{y_\alpha : \omega \le \alpha < \mu\}$ be a dense subset in Y. We shall proceed by transfinite induction on α. Put $A_0 = \{y_\omega\}$, $B_0 = \emptyset$, $\aleph = \aleph_0$ in Lemma 6.1.3; we thus find sets $A_0 \subset A \subset V$, $B_0 \subset B \subset W$ and we denote $A_{\omega+1} = A$, $B_{\omega+1} = B$.

Let $\omega + 1 < \gamma \le \mu$ be fixed and assume that for every $\omega < \alpha < \gamma$ we have already constructed the sets $A_\alpha \subset V$ and $B_\alpha \subset W$ satisfying the properties listed in the lemma as well as the condition that

$$f(A_\beta) \subset A_\alpha \quad \text{if} \quad \omega < \beta < \alpha.$$

If γ is a limit ordinal, simply put $A_\gamma = \bigcup_{\alpha < \gamma} A_\alpha$ and $B_\gamma = \bigcup_{\alpha < \gamma} B_\alpha$. Next, assume γ is a nonlimit ordinal. By applying Lemma 6.1.3 for $A_0 = A_{\gamma-1} \cup \{y_{\gamma-1}\} \cup f(A_{\gamma-1})$, $B_0 = B_{\gamma-1}$, and $\aleph = $ card γ, we obtain sets A_γ, B_γ. Then it is easy to verify that both A_γ and B_γ share all the properties listed above, perhaps except the identity $\overline{A_\gamma} \cap Y = \overline{A_\gamma \cap Y}$ in the case when γ is a limit ordinal. Let us check that this also holds. Take a sequence $\{a_i\}$ in A_γ converging to some $y \in Y$. Then

$$\|f(a_i) - y\| \le 3\|a_i - y\| \to 0$$

and $f(a_i) \in f(A_{\beta_i}) \subset A_{\beta_i+1} \subset A_\gamma$ for some $\beta_i < \gamma$. So $y \in \overline{A_\gamma \cap Y}$ and $\overline{A_\gamma} \cap Y = \overline{A_\gamma \cap Y}$. Finally, since A_μ contains the set $\{y_\alpha : \omega \le \alpha < \mu\}$, which is dense in Y, we conclude that $\overline{A_\mu} \supset Y$. $\qquad \square$

Definition 6.1.5. Let V be a nonseparable Banach space and let μ be the first ordinal with card $\mu = $ dens V. A *projectional resolution of the identity* (*P.R.I.*) on V is a "long sequence", that is, a family $\{P_\alpha : \omega \le \alpha \le \mu\}$ of linear projections on V such that $P_\omega \equiv 0$, P_μ is the identity mapping, and for all $\omega < \alpha \le \mu$ the following hold:

(i) $\|P_\alpha\| = 1$,

(ii) dens $P_\alpha V \leq$ card α,

(iii) $P_\alpha \circ P_\beta = P_\beta \circ P_\alpha = P_\beta$ if $\omega \leq \beta \leq \alpha$, and

(iv) $\bigcup_{\beta < \alpha} P_{\beta+1} V$ is norm-dense in $P_\alpha V$.

In some important, at least for our purposes, classes of Banach spaces, the following concept can be traced behind the existence of a P.R.I.:

Definition 6.1.6. Let V be a Banach space and W be a one-norming subset of V^*, that is, for every $v \in V$ we have $\|v\| = \sup\langle B_{V^*} \cap W, v\rangle$, and let us assume that $\overline{W}$ is linear. Let $\Phi : W \to 2^V$ be an at most countably valued mapping such that for every nonempty set $B \subset W$, with $\overline{B}$ linear,

$$\Phi(B)^\perp \cap \overline{B}^* = \{0\}.$$

Then the couple (W, Φ) is called a *projectional generator* on V.

Proposition 6.1.7.

A nonseparable Banach space with a projectional generator admits a P.R.I.

Proof. Let (W, Φ) be a projectional generator on a Banach space V. Put $P_\omega \equiv 0$. For every $v \in V$ we find a countable set $\Psi(v)$ in W such that

$$\|v\| = \sup\langle \Psi(v) \cap B_{V^*}, v\rangle;$$

this is possible because W is a one-norming set. Thus, we have defined an at most countably valued mapping $\Psi : V \to 2^W$. Now, applying Proposition 6.1.4 with $Y = V$, we get "long sequences" $\{A_\alpha : \omega < \alpha \leq \mu\}$ and $\{B_\alpha : \omega < \alpha \leq \mu\}$ of sets belonging to V and W, respectively. Fix any $\omega < \alpha \leq \mu$. Then we have, for every $a \in A_\alpha$,

$$\|a\| = \sup\langle \Psi(a) \cap B_{V^*}, a\rangle \leq \sup\langle B_\alpha \cap B_{V^*}, a\rangle \leq \|a\|$$

as $\Psi(A_\alpha) \subset B_\alpha$, and

$$A_\alpha^\perp \cap \overline{B_\alpha}^* \subset \Phi(B_\alpha)^\perp \cap \overline{B_\alpha}^* = \{0\}$$

as $\Phi(B_\alpha) \subset A_\alpha$ and (W, Φ) is a projectional generator. It then follows, using Lemma 6.1.1, that there exists a linear norm 1 projection $P_\alpha : V \to V$ such that $P_\alpha V = \overline{A_\alpha}$, $P_\alpha^{-1}(0) = B_{\alpha\perp}$, and $P_\alpha^* V^* = \overline{B_\alpha}^*$. Performing this for every $\alpha \in (\omega, \mu]$, the properties of $\{A_\alpha\}$ and $\{B_\alpha\}$ listed in Proposition 6.1.4 immediately yield that $\{P_\alpha : \omega \leq \alpha \leq \mu\}$ is a P.R.I. on V. $\square$

Remark 6.1.8. From the above proof, we can see that, according to Lemma 6.1.2, the requirement $\Phi(B)^\perp \cap \overline{B}^* = \{0\}$ can be replaced by the weaker one $\Phi(B)^\perp \cap \overline{B \cap B_{V^*}}^* = \{0\}$ in the definition of a projectional generator. This will be of use in the proofs of Propositions 8.2.1 and 8.3.1.

In Chapter 8, the following assertion will be useful:

Proposition 6.1.9.

Suppose that a nonseparable dual Banach space V^* admit a projectional generator (W, Φ) such that $W \subset V$ and $\overline{W} = V$. Then V^* admits a P.R.I. $\{P_\alpha : \omega \le \alpha \le \mu\}$ together with a nondecreasing "long sequence" $\{E_\alpha : \omega \le \alpha \le \mu\}$ of subspaces of V such that $E_\omega = \{0\}$ and for all $\omega < \alpha \le \mu$ the following hold:

(i) $\dim E_\alpha \le \operatorname{card} \alpha$,
(ii) $E_\alpha = \overline{\bigcup_{\beta < \alpha} E_{\beta+1}}$,
(iii) the mapping R_α sending $\xi \in P_\alpha V^*$ to its restriction $\xi_{|E_\alpha}$ maps $P_\alpha V^*$ isometrically onto E_α^*, and
(iv) $(P_{\alpha+1} - P_\alpha)V^*$ is isometric to $(E_{\alpha+1}/E_\alpha)^*$ if $\alpha < \mu$.

Proof. Put $E_\omega = \{0\}$ and let μ be the first ordinal with card $\mu = \dim V^*$. We shall repeat, almost word by word, the proof of Proposition 6.1.7. Thus, for each $\alpha \in (\omega, \mu]$ we obtain sets $A_\alpha \subset V^*$, $B_\alpha \subset W$ ($\subset V \subset V^{**}$) and projections $P_\alpha : V^* \to V^*$ such that $P_\alpha V^* = \overline{A_\alpha}$, $P_\alpha^{-1}(0) = B_\alpha^\perp$, $P_\alpha^* V^{**} = \overline{B_\alpha}^*$ and $\|a\| = \sup\langle a, B_\alpha \cap B_V \rangle$ for all $a \in A_\alpha$. Then $\{P_\alpha : \omega \le \alpha \le \mu\}$ is a P.R.I. on V^*.

Put $E_\alpha = \overline{B_\alpha}$; then (i) and (ii) hold. Define the mappings $R_\alpha : P_\alpha V^* \to E_\alpha^*$ by $R_\alpha \xi = \xi_{|E_\alpha}$, $\xi \in P_\alpha V^*$. Thus, each R_α is an isometric embedding. Now take any $\eta \in E_\alpha^*$ and let $\xi \in V^*$ be such that $\xi_{|E_\alpha} = \eta$. Then for all $v \in E_\alpha$ we have $\langle P_\alpha \xi, v \rangle = \langle \eta, v \rangle$, since $P_\alpha \xi - \xi \in B_\alpha^\perp$. Hence $R_\alpha(P_\alpha \xi) = \eta$, which means that R_α is surjective.

As to the proof of (iv), we define a mapping $\varphi : (P_{\alpha+1} - P_\alpha)V^* \to (E_{\alpha+1}/E_\alpha)^*$ by

$$\langle \varphi(\xi), [v] \rangle = \langle \xi, v \rangle, \qquad \xi \in (P_{\alpha+1} - P_\alpha)V^*, \qquad [v] \in E_{\alpha+1}/E_\alpha;$$

here $[v]$ means the class $v + E_\alpha$. It is well defined since for $\xi \in (P_{\alpha+1} - P_\alpha)V^*$ and for $v \in E_\alpha$ we have

$$\langle \xi, v \rangle = \langle (P_{\alpha+1} - P_\alpha)\xi, v \rangle = \langle P_{\alpha+1}^* v, \xi \rangle - \langle P_\alpha^* v, \xi \rangle = 0$$

because $P_\alpha^* V^{**} = \overline{E_\alpha}^*$. Moreover, for every $\xi \in (P_{\alpha+1} - P_\alpha)V^*$ we get

$$\|\varphi(\xi)\| = \sup\left\{\langle\varphi(\xi), [v]\rangle : \ [v] \in E_{\alpha+1}/E_\alpha, \ \|[v]\| < 1\right\}$$
$$= \sup\left\{\langle\xi, v\rangle : \ v \in E_{\alpha+1}, \ \|v\| < 1\right\} = \|R_{\alpha+1}\xi\| = \|\xi\|$$

as $(P_{\alpha+1} - P_\alpha)V^* \subset P_{\alpha+1}V^*$, and we already know that $R_{\alpha+1}$ is an isometry. It remains to prove that φ is onto. Let $\eta \in (E_{\alpha+1}/E_\alpha)^*$ be given and define $\zeta \in E_{\alpha+1}^*$ by $\langle\zeta, v\rangle = \langle\eta, [v]\rangle$, $v \in E_{\alpha+1}$. Also put $\xi = R_{\alpha+1}^{-1}(\zeta)$. Take any $v \in V$. Then $P_\alpha^* v \in P_\alpha^* V^{**} = \overline{B_\alpha}^*$. Let $\{u_\tau\} \subset B_\alpha$ be a net which is weak* convergent to $P_\alpha^* v$. We then have

$$\langle P_\alpha\xi, v\rangle = \langle P_\alpha^* v, \xi\rangle = \lim_\tau\langle\xi, u_\tau\rangle$$
$$= \lim_\tau\langle\zeta, u_\tau\rangle = \lim_\tau\langle\eta, [u_\tau]\rangle = 0$$

since $u_\tau \in B_\alpha \subset E_\alpha$, and so $P_\alpha\xi = 0$. Hence, for all $[v] \in E_{\alpha+1}/E_\alpha$ we get

$$\langle\varphi((P_{\alpha+1} - P_\alpha)\xi), [v]\rangle = \langle((P_{\alpha+1} - P_\alpha)\xi), v\rangle$$
$$= \langle P_{\alpha+1}\xi, v\rangle = \langle\xi, v\rangle = \langle\zeta, v\rangle = \langle\eta, [v]\rangle;$$

that is, $\varphi((P_{\alpha+1} - P_\alpha)\xi) = \eta$, which means that φ is surjective. $\qquad\square$

Let us remark that, *under the hypotheses of Proposition 6.1.9, V is an Asplund space.* In fact, let Y be a separable subspace of V. We find a countable set $B \subset W$ with $\overline{B} \supset Y$. We shall show that $\overline{\mathrm{sp}}\{\xi_{|Y} : \ \xi \in \Phi(B)\} = Y^*$, and so Y^* will be separable. Thus, according to Theorem 1.1.1, we shall be done. Assume this is not true. Then there is $\emptyset \neq y^{**} \in Y^{**}$ such that $\langle y^{**}, \xi_{|Y}\rangle = 0$ whenever $\xi \in \Phi(B)$. Define $v^{**} \in V^{**}$ by $\langle v^{**}, v^*\rangle = \langle y^{**}, v^*_{|Y}\rangle$, $v^* \in V^*$. Then $v^{**} \in \Phi(B)^\perp$. On the other hand, y^{**} is, by Goldstine's theorem, a weak* limit of some bounded net $\{y_\tau\} \subset Y$. Let u^{**} be a weak* cluster point of $\{y_\tau\}$ in V^{**}. Then necessarily $u^{**} = v^{**}$. Hence, $v^{**} \in \overline{Y}^* \subset \overline{B}^*$. Since (W, Φ) is a projectional generator, it follows that $v^{**} = 0$ and so $y^{**} = 0$, a contradiction. For the converse statement see Proposition 8.2.1.

For an easier understanding of the proof of the next proposition, the reader may first consider the special case when $Y = V$; then the business with the norms $\|\cdot\|_n$, $n = 1, 2, \ldots$, is superfluous. The general case, when $Y \subset V$, will be used in the proofs of Proposition 8.3.2 and Theorem 8.3.4. However, it is also possible to prove these statements from the special case $Y = V$ once we know that the continuous image of a Corson compact is a Corson compact; see Section 8.3 for details.

Proposition 6.1.10.

Let a nonseparable Banach space V have a projectional generator (W, Φ) with $W = V^*$. Let Y be a nonseparable subspace of V, let μ be the first ordinal with card $\mu = $ dens Y, and assume that the dual Y^* has a projectional generator

(Z, Ψ) with $Z = Y$. Then there exist linear projections $R_\alpha : V \to V$, $\omega \le \alpha \le \mu$ satisfying (i)–(iv) in Definition 6.1.5 and such that, when denoting $P_\alpha = R_{\alpha|Y}$, then $\{P_\alpha : \omega \le \alpha \le \mu\}$ is a P.R.I. on Y and $\{P_\alpha^* : \omega \le \alpha \le \mu\}$ is a P.R.I. on Y^*.

Proof. Denote by $\| \cdot \|_0$ the original norm on V. If $v^* \in V^*$, let $\widetilde{\Phi}(v^*)$ be a countable subset of V such that $\widetilde{\Phi}(v^*) \supset \Phi(v^*)$ and

$$\|v^*_{|Y}\|_0 = \sup\langle v^*, \widetilde{\Phi}(v^*) \cap B_Y\rangle.$$

Note that $(V^*, \widetilde{\Phi})$ will also be a projectional generator on V. For $n = 1, 2, \ldots$ let $\| \cdot \|_n$ be the (equivalent) norm on V whose unit ball is the closure of $B_Y + (1/n)B_V$. (Here $B_Y = \{y \in Y : \|y\|_0 \le 1\}$.) For $n = 0, 1, 2, \ldots$ let V_n denote the space $(V, \| \cdot \|_n)$. For $v \in V$, let $\widetilde{\Psi}(v)$ be a countable subset of V^* such that, for all $n = 0, 1, 2, \ldots$,

$$\|v\|_n = \sup \langle \widetilde{\Psi}(v) \cap B_{V_n^*}, v\rangle,$$

and

$$\left\{ v^*_{|Y} : v^* \in \widetilde{\Psi}(v) \right\} \supset \Psi(v) \quad \text{for each} \quad v \in Y.$$

Now, for the mappings $\widetilde{\Phi}$ and $\widetilde{\Psi}$, we shall perform the same process as done in Proposition 6.1.4. We shall thus obtain "long sequences" $\{A_\alpha : \omega < \alpha \le \mu\}$ and $\{B_\alpha : \omega < \alpha \le \mu\}$ of sets in V and V^*, respectively. Let us recall that, among other properties, the sets A_α and B_α satisfy the inclusions $\widetilde{\Phi}(B_\alpha) \subset A_\alpha$ and $\widetilde{\Psi}(A_\alpha) \subset B_\alpha$. Then, almost exactly as in the proof of Proposition 6.1.7, we can find a "long sequence" $\{R_\alpha : \omega \le \alpha \le \mu\}$ of linear projections on V such that $R_\omega \equiv 0$ and for all $\omega < \alpha \le \mu$ we have $\|R_\alpha\|_n = 1$, $n = 0, 1, \ldots$, $R_\alpha \circ R_\beta = R_{\min(\alpha,\beta)}$, $\operatorname{dens} R_\alpha V \le \operatorname{card}\alpha$, $R_\alpha V = \bigcup_{\beta < \alpha} R_{\beta+1}V$, $R_\alpha V = \overline{A_\alpha}$, and $R_\alpha^{-1}(0) = B_{\alpha\perp}$.

Now put $P_\alpha = R_{\alpha|Y}$, $\omega \le \alpha \le \mu$. As $\|R_\alpha\|_n = 1$, R_α maps $B_Y + (1/n)B_V$ into its closure; hence

$$R_\alpha(B_Y) \subset \bigcap_{n=1}^{\infty} \overline{B_Y + \frac{1}{n}B_V} \subset B_Y$$

and so $\|P_\alpha\|_0 = 1$ and $P_\alpha Y \subset Y$ for every $\omega < \alpha \le \mu$. Trivially, $P_\alpha \circ P_\beta = P_{\min (\alpha,\beta)}$ and $\operatorname{dens} P_\alpha Y \le \operatorname{dens} R_\alpha V \le \operatorname{card}\alpha$. As regards to (iv) in Definition 6.1.5, take a limit $\omega < \alpha \le \mu$, $y \in P_\alpha Y$, and $\epsilon > 0$. We find $\omega < \beta < \alpha$ and $v \in R_\beta V$ such that $\|y - v\| < \epsilon/2$. Then $\|v - P_\beta y\| = \|R_\beta(v - y)\| \le \|v - y\| < \epsilon/2$ and so $\|y - P_\beta y\| \le \|y - v\| + \|v - P_\beta y\| < \epsilon$, which proves (iv). Finally, recall that $P_\mu Y = R_\mu Y = \overline{A_\mu} \supset Y$. Therefore $\{P_\alpha : \omega \le \alpha \le \mu\}$ is a P.R.I. on Y.

Now take any $\omega < \alpha \le \mu$ and put $C_\alpha = A_\alpha \cap Y$, $D_\alpha = \{v^*_{|Y} : v^* \in B_\alpha\}$. We choose $y^* \in D_\alpha$ and we find $v^* \in B_\alpha$ so that $v^*_{|Y} = y^*$. Then

$$\|y^*\|_0 = \|v^*{}_{|Y}\|_0 = \sup\langle v^*, \widetilde{\Phi}(v^*) \cap B_Y\rangle$$
$$\leq \sup\langle v^*, A_\alpha \cap B_Y\rangle = \sup\langle v^*, C_\alpha \cap B_Y\rangle \leq \|y^*\|_0,$$

as $\widetilde{\Phi}(B_\alpha) \subset A_\alpha$. For $M \subset V^*$, put $M_{|Y} = \{v^*{}_{|Y} : v^* \in M\}$. Then

$$\Psi(C_\alpha) \subset \widetilde{\Psi}(C_\alpha)_{|Y} \subset \widetilde{\Psi}(A_\alpha)_{|Y} \subset B_{\alpha|Y} = D_\alpha,$$

as $\widetilde{\Psi}(A_\alpha) \subset B_\alpha$. So $\overline{D_\alpha}^\perp \cap \overline{C_\alpha}^* \subset \Psi(C_\alpha)^\perp \cap \overline{C_\alpha}^* = \{0\}$ since (Y, Ψ) is a projectional generator on Y^*. Now recall that $\overline{A_\alpha}$, $\overline{B_\alpha}$ are linear. Then $\overline{C_\alpha} = \overline{A_\alpha} \cap Y$ by Lemma 6.1.4 and so $\overline{C_\alpha}$ is linear. The set $\overline{D_\alpha}$ also is linear. It then follows from Lemma 6.1.1 that there exists, on Y^*, a linear norm 1 projection $Q_\alpha : Y^* \to Y^*$ such that $Q_\alpha Y^* = \overline{D_\alpha}$, $Q_\alpha^* Y^{**} = \overline{C_\alpha}^*$, and $Q_\alpha^{-1}(0) = C_\alpha^\perp$. Put $Q_\omega \equiv 0$. Using Lemma 6.1.4, we can easily check that $\{Q_\alpha\}$ is a P.R.I. on Y^*.

Next, for $y \in Y$ and $y^* \in Y^*$, we have

$$\langle Q_\alpha y^*, y\rangle = \langle Q_\alpha y^*, y - P_\alpha y\rangle + \langle Q_\alpha y^*, P_\alpha y\rangle = \langle Q_\alpha y^*, P_\alpha y\rangle$$

as $Q_\alpha y^* \in \overline{D_\alpha}$ and $y - P_\alpha y \in R_\alpha^{-1}(0) = B_{\alpha\perp}$. We also have $y^* - Q_\alpha y^* \in Q_\alpha^{-1}(0) = C_\alpha^\perp$ and $P_\alpha y \in \overline{A_\alpha} \cap Y$. However, by Proposition 6.1.4, $\overline{A_\alpha} \cap Y = A_\alpha \cap Y$. So $P_\alpha y \in \overline{C_\alpha}$ and thus

$$\langle y^*, P_\alpha y\rangle = \langle y^* - Q_\alpha y^*, P_\alpha y\rangle + \langle Q_\alpha y^*, P_\alpha y\rangle = \langle Q_\alpha y^*, P_\alpha y\rangle = \langle Q_\alpha y^*, y\rangle.$$

It follows that $P_\alpha^* = Q_\alpha$ for all α and hence $\{P_\alpha^*\}$ is a P.R.I on Y^*. □

Let us note that if $\{P_\alpha\}$ is a P.R.I. on V, then $\{P_\alpha^*\}$ may not be a P.R.I. on V^*. Indeed, the density of $P_\alpha^* V^*$ may exceed card α, and further, $\bigcup_{\beta<\alpha} P_\beta^* V^*$ is in general far from being dense in $P_\alpha^* V^*$.

6.2. CONSEQUENCES OF THE EXISTENCE OF A P.R.I.

Proposition 6.2.1.

Let $\{P_\alpha : \omega \leq \alpha \leq \mu\}$ be a P.R.I. on a nonseparable Banach space V and take any $v \in V$. Then
  (i) the mapping $\beta \mapsto P_\beta v$ from $[\omega, \mu]$, endowed with the order topology, into V with the norm topology is continuous,
  (ii) for every $\epsilon > 0$ the set $\{\omega \leq \alpha < \mu : \|(P_{\alpha+1} - P_\alpha)v\| > \epsilon\}$ is finite,
  (iii) v lies in the set

$$\Lambda(v) := \overline{sp}\{(P_{\alpha+1} - P_\alpha)v : \omega \leq \alpha < \mu\},$$

  (iv) $V = \overline{sp}\left(\bigcup\{(P_{\alpha+1} - P_\alpha)V : \omega \leq \alpha < \mu\}\right).$

Proof. (i) Let a limit ordinal $\omega < \alpha \leq \mu$ and $\epsilon > 0$ be given. By (iv) in Definition 6.1.5, there are $\omega \leq \gamma < \alpha$ and $u \in P_\gamma V$ so that $\|P_\alpha v - u\| < \epsilon/2$. Then for all $\gamma \leq \beta \leq \alpha$ we have

$$\|P_\beta v - u\| = \|P_\beta(P_\alpha v - u)\| \leq \|P_\alpha v - u\| < \frac{\epsilon}{2}.$$

So

$$\|P_\alpha v - P_\beta v\| \leq \|P_\alpha v - u\| + \|u - P_\beta v\| < 2\frac{\epsilon}{2} = \epsilon.$$

(ii) In order to obtain a contradiction, assume there is $\epsilon > 0$ and an infinite sequence $\{\alpha_i\} \subset [\omega, \mu)$ such that

$$\|(P_{\alpha_{i+1}} - P_{\alpha_i})\| > \epsilon \quad \text{for all} \quad i = 1, 2, \dots .$$

By going to a subsequence, if necessary, we may and do assume that $\omega \leq \alpha_1 < \alpha_2 < \cdots < \mu$. Put $\alpha = \lim_i \alpha_i$. Then, by (i), for all sufficiently large $i \in \mathbb{N}$ we have

$$\|P_{\alpha_i} v - P_\alpha v\| < \epsilon/2, \qquad \|P_{\alpha_{i+1}} v - P_\alpha v\| < \epsilon/2,$$

so $\|P_{\alpha_{i+1}} v - P_{\alpha_i} v\| < \epsilon$, which gives us a contradiction.

(iii) Take any $v \in V$. Trivially, $P_\omega v = 0 \in \Lambda(v)$. Assume we have $\omega < \gamma \leq \mu$ such that $P_\alpha v \in \Lambda(v)$ whenever $\alpha < \gamma$. If γ is a limit ordinal, then (i) guarantees that $P_\gamma v \in \Lambda(v)$ as well. If γ is a non-limit ordinal, then

$$P_\gamma v = (P_\gamma - P_{\gamma-1})v + P_{\gamma-1} v \in \Lambda(v) + \Lambda(v) = \Lambda(v).$$

(iv) This follows immediately from (iii). $\qquad\square$

Proposition 6.2.2.

Suppose that a nonseparable Banach space V admits a P.R.I. $\{P_\alpha : \omega \leq \alpha \leq \mu\}$ and assume that for every $\omega \leq \alpha < \mu$ there exist a set Γ_α and a bounded linear one-to-one mapping T_α from $(P_{\alpha+1} - P_\alpha)V$ into $c_0(\Gamma_\alpha)$. Then there exist a set Γ and a bounded linear one-to-one mapping from V into $c_0(\Gamma)$.

Proof. Let Γ denote a "disjoint" union of all the Γ_α (i.e., we arrange things in such a way that $\Gamma_\alpha \cap \Gamma_\beta = \emptyset$ if $\alpha \neq \beta$), where $\omega \leq \alpha < \mu$, and define a mapping $T : V \to \mathbb{R}^\Gamma$ by

$$Tv(\gamma) = \|T_\alpha\|^{-1} T_\alpha\big((P_{\alpha+1} - P_\alpha)v\big)(\gamma) \quad \text{if} \quad \gamma \in \Gamma_\alpha, \qquad \omega \leq \alpha < \mu; \qquad v \in V.$$

Clearly, T is linear and $|Tv(\gamma)| \leq 2\|v\|$ for all $v \in V$ and all $\gamma \in \Gamma$. Choose $v \in V$ so that $Tv = 0$. Since all the T_α are injective, we get that $P_{\alpha+1} v = P_\alpha v$ for all $\omega \leq \alpha < \mu$. This means, by Proposition 6.2.1(iii), that $v = 0$.

It remains to show that $Tv \in c_0(\Gamma)$. So take any $v \in V$ and any $\epsilon > 0$. By Proposition 6.2.1(ii), there is a finite set $F \subset [\omega, \mu)$ such that

$$\left\|(P_{\alpha+1} - P_\alpha)v\right\| < \epsilon \quad \text{whenever} \quad \alpha \in [\omega, \mu)\backslash F;$$

hence $|Tv(\gamma)| < \epsilon$ for all $\gamma \in \bigcup\{\Gamma_\alpha : \alpha \in [\omega, \mu)\backslash F\}$. Further, for every $\alpha \in \Gamma$ and, in particular, for every $\alpha \in F$, the set

$$\left\{\gamma \in \Gamma_\alpha : \left|T_\alpha((P_{\alpha+1} - P_\alpha)v)(\gamma)\right| > \epsilon\|T_\alpha\|\right\}$$

is finite since the range of T_α is in $c_0(\Gamma_\alpha)$. Therefore, $\{\gamma \in \Gamma : |Tv(\gamma)| > \epsilon\}$ must be a finite set. We have thus shown that Tv lies in $c_0(\Gamma)$. $\qquad\square$

Definition 6.2.3. Let V be a Banach space. A system $\{(v_\gamma, v_\gamma^*) \in V \times V^* : \gamma \in \Gamma\}$ is called a *Markuševič basis* in V if $\langle v_\beta^*, v_\gamma \rangle = \delta_{\beta\gamma}$ (the Kronecker delta) for all $\beta, \gamma \in \Gamma$, if $\overline{\mathrm{sp}}\{v_\gamma : \gamma \in \Gamma\} = V$, and if the set $\{v_\gamma^* : \gamma \in \Gamma\}$ is total on V, that is, if for every nonzero $v \in V$ there is a $\gamma \in \Gamma$ such that $\langle v_\gamma^*, v \rangle \neq 0$. If moreover $\overline{\mathrm{sp}}\{v_\gamma^* : \gamma \in \Gamma\} = V^*$, then we speak about a *shrinking Markuševič basis* in V.

It should be mentioned here that *a Banach space with a shrinking Markuševič basis* $\{(v_\gamma, v_\gamma^*) : \gamma \in \Gamma\}$ *is WCG and Asplund*. Indeed, the set $\{\|v_\gamma\|^{-1}v_\gamma : \gamma \in \Gamma\} \cup \{0\}$ is weakly compact and linearly dense in V. Further, let Y be a separable subspace of V. Then there is a countable set $\Gamma_0 \subset \Gamma$ such that $Y \subset \overline{\mathrm{sp}}\{v_\gamma : \gamma \in \Gamma_0\}$ and we can easily verify that $Y^* \subset \overline{\mathrm{sp}}\{v_{\gamma|Y}^* : \gamma \in \Gamma_0\}$. Indeed, take any $y^* \in Y^*$ and any $\epsilon > 0$. Let $v^* \in V^*$ be such that $v^*|_Y = y^*$. We may find a finite set $F \subset \Gamma$ and $a_\gamma \in \mathbb{R}$, $\gamma \in F$, such that $\|v^* - \sum_{\gamma \in F} a_\gamma v_\gamma^*\| < \epsilon$. Then $\|y^* - \sum_{\gamma \in F \cap \Gamma_0} a_\gamma v_{\gamma|Y}^*\| < \epsilon$. Therefore, Y^* is separable, and so Theorem 1.1.1 ensures that V is Asplund. For the converse statement see Theorem 8.3.3.

Proposition 6.2.4.

Suppose that a nonseparable Banach space V has a P.R.I. $\{P_\alpha : \omega \leq \alpha \leq \mu\}$ and assume that for every $\omega \leq \alpha < \mu$ the space $(P_{\alpha+1} - P_\alpha)V$ has a Markuševič basis. Then V has a Markuševič basis.

Proof. For $\omega \leq \alpha < \mu$ let $\{(v_{\alpha\gamma}, v_{\alpha\gamma}^*) : \gamma \in \Gamma_\alpha\}$ be a Markuševič basis for $(P_{\alpha+1} - P_\alpha)V$ and define the mapping $Q_\alpha : V \to (P_{\alpha+1} - P_\alpha)V$ by $Q_\alpha v = (P_{\alpha+1} - P_\alpha)v$, $v \in V$. We shall show that

$$\left\{(v_{\alpha\gamma}, Q_\alpha^* v_{\alpha\gamma}^*) : \gamma \in \Gamma_\alpha, \omega \leq \alpha < \mu\right\}$$

is a Markuševič basis for V. Using the commutativity of $\{P_\alpha\}$, we can easily verify that

$$\langle Q^*_\alpha v^*_{\alpha\gamma}, v_{\alpha'\gamma'} \rangle = \begin{cases} 1 & \text{if } \alpha = \alpha' \text{ and } \gamma = \gamma', \\ 0 & \text{otherwise.} \end{cases}$$

Proposition 6.2.1(iv) immediately gives that

$$\overline{\mathrm{sp}}\{v_{\alpha\gamma} : \gamma \in \Gamma_\alpha, \ \omega \le \alpha < \mu\} = V.$$

Further, let $0 \ne v \in V$ be given. Applying Proposition 6.2.1(iii), we get that $(P_{\alpha+1} - P_\alpha)v \ne 0$ for some $\alpha \in [\omega, \mu)$. Hence

$$\langle Q^*_\alpha v^*_{\alpha\gamma}, v \rangle = \langle v^*_{\alpha\gamma}, (P_{\alpha+1} - P_\alpha)v \rangle \ne 0$$

for some $\gamma \in \Gamma_\alpha$. $\qquad\square$

Proposition 6.2.5.

Let V be a nonseparable Banach space with a P.R.I. $\{P_\alpha : \omega \le \alpha \le \mu\}$ such that $\{P^*_\alpha : \omega \le \alpha \le \mu\}$ is a P.R.I. on V^*.
  (i) If for each $\omega \le \alpha < \mu$ the subspace $(P_{\alpha+1} - P_\alpha)V$ is WCG, then so is V.
  (ii) If for each $\omega \le \alpha < \mu$ the subspace $(P_{\alpha+1} - P_\alpha)V$ has a shrinking Markuševič basis, then so does V.

Proof. (i) For $\omega \le \alpha < \mu$ let K_α be a weakly compact set such that $\overline{\mathrm{sp}}\,K_\alpha = (P_{\alpha+1} - P_\alpha)V$. We may and do assume that $K_\alpha \subset B_V$ for every α. Define

$$K = \bigcup \{K_\alpha : \omega \le \alpha < \mu\} \cup \{0\}.$$

Proposition 6.2.1(iv) ensures that K is linearly dense in V. It remains to show that K is weakly compact. Let $\{k_\tau\}$ be a net in K. If there exists a subnet of $\{k_\tau\}$ which lies in K_α for some fixed α, we are done. Otherwise for any fixed $\alpha \in [\omega, \mu)$ we have $k_\tau \notin K_\alpha$ for all large τ. Hence, using "orthogonality" of $\{P_{\alpha+1} - P_\alpha\}$, we get that, for every $v^* \in V^*$ and every $\alpha \in [\omega, \mu)$,

$$\lim_\tau \left\langle (P^*_{\alpha+1} - P^*_\alpha)v^*, k_\tau \right\rangle = \lim_\tau \left\langle v^*, (P_{\alpha+1} - P_\alpha)k_\tau \right\rangle = \lim_\tau \langle v^*, 0 \rangle = 0.$$

Now Proposition 6.2.1(iii) applied to $\{P^*_\alpha\}$ guarantees that $\{k_\tau\}$ converges weakly to 0.

  (ii) Consider the mappings Q_α defined as in the proof of Proposition 6.2.4. Then for each $\xi \in \left((P_{\alpha+1} - P_\alpha)V\right)^*$ we have

$$2\|\xi\| \ge \|Q^*_\alpha \xi\| = \sup\langle Q^*_\alpha \xi, B_V \rangle = \sup\langle \xi, Q_\alpha(B_V) \rangle$$
$$= \sup\langle \xi, (P_{\alpha+1} - P_\alpha)(B_V) \rangle \ge \sup\langle \xi, B_{(P_{\alpha+1}-P_\alpha)V} \rangle = \|\xi\|,$$

and so Q_α^* is an isomorphic embedding. It is easy to check that the range of Q_α^* lies in $(P_{\alpha+1}^* - P_\alpha^*)V^*$. Further, take any $\eta \in (P_{\alpha+1}^* - P_\alpha^*)V^*$ and define $\xi(v) = \langle \eta, v \rangle$ for $v \in (P_{\alpha+1} - P_\alpha)V$; so $\xi \in ((P_{\alpha+1} - P_\alpha)V)^*$. Then for all $v \in V$ we get

$$\langle Q_\alpha^* \xi, v \rangle = \langle \xi, (P_{\alpha+1} - P_\alpha)v \rangle = \langle \eta, (P_{\alpha+1} - P_\alpha)v \rangle = \langle \eta, v \rangle$$

so that $Q_\alpha^* \xi = \eta$ and hence the range of Q_α^* is $(P_{\alpha+1}^* - P_\alpha^*)V^*$. For $\omega \le \alpha < \mu$ let $\{(v_{\alpha\gamma}, v_{\alpha\gamma}^*) : \gamma \in \Gamma_\alpha\}$ be a shrinking Markuševič basis for $(P_{\alpha+1} - P_\alpha)V$. From Proposition 6.2.4, we already know that $\{(v_{\alpha\gamma}, Q_\alpha^* v_{\alpha\gamma}^*) : \gamma \in \Gamma_\alpha, \omega \le \alpha < \mu\}$ is a Markuševič basis in V. It remains to show that it is shrinking. Since $\{P_\alpha^*\}$ is a P.R.I. on V^*, we know from Proposition 6.2.1(iv) that

$$\overline{\mathrm{sp}}\left(\bigcup \{(P_{\alpha+1}^* - P_\alpha^*)V^* : \omega \le \alpha < \mu\}\right) = V^*.$$

Therefore

$$\overline{\mathrm{sp}}\{Q_\alpha^* v_{\alpha\gamma}^* : \gamma \in \Gamma_\alpha, \omega \le \alpha < \mu\} = V^*. \qquad \square$$

Sometimes the following variant of the P.R.I. is useful.

Definition 6.2.6. Let V be a nonseparable Banach space and let μ be the first ordinal with card $\mu = $ dens V. A *separable projectional resolution of the identity* (S.P.R.I.) on V is a "long sequence" $\{Q_\alpha : \omega \le \alpha \le \mu\}$ of linear projections on V such that $Q_\omega \equiv 0$, Q_μ is the identity mapping, and for each $\omega \le \alpha < \mu$ the following holds:
  (i) $\|Q_\alpha\| < +\infty$,
  (ii) $(Q_{\alpha+1} - Q_\alpha)V$ is separable,
  (iii) $Q_\alpha \circ Q_\beta = Q_\beta \circ Q_\alpha = Q_\beta$ if $\omega \le \beta \le \alpha$, and
  (iv) $v \in \overline{\mathrm{sp}}\{(Q_{\alpha+1} - Q_\alpha)v : \omega \le \alpha < \mu\}$ for all $v \in V$.

Clearly, by Proposition 6.2.1(iii), if a Banach space has density $\aleph_1$, then every P.R.I. on it is also a S.P.R.I.

Proposition 6.2.7.

Let V be a nonseparable Banach space with a P.R.I. $\{P_\alpha : \omega \le \alpha \le \mu\}$ and assume that for every $\omega \le \alpha < \mu$ the space $(P_{\alpha+1} - P_\alpha)V$ is either separable or admits a S.P.R.I. Then the whole space V admits a S.P.R.I.

Proof. Take any $\alpha \in [\omega, \mu)$. If $(P_{\alpha+1} - P_\alpha)V$ is separable, put $\mu_\alpha = 2\omega$, $Q_\omega^\alpha \equiv 0$, and $Q_\beta^\alpha = P_{\alpha+1} - P_\alpha$ for all $\omega < \beta < 2\omega$. If $(P_{\alpha+1} - P_\alpha)V$ is nonseparable, let $\{Q_\beta^\alpha : \omega \le \beta \le \mu_\alpha\}$ be a S.P.R.I. on the subspace $(P_{\alpha+1} - P_\alpha)V$; so card $\mu_\alpha \le$ dens $P_{\alpha+1}V \le$ card α. Define the set

$$\Lambda = \big\{(\alpha,\beta):\ \omega \le \beta < \mu_\alpha,\ \omega \le \alpha < \mu\big\} \cup \big\{(\mu,\omega)\big\}$$

and endow it with the lexicographical order, that is, $(\alpha,\beta) \le (\alpha',\beta')$ if and only if either $\alpha < \alpha'$ or $\alpha = \alpha'$ and $\beta \le \beta'$; this will be a well-ordered set. We observe that for every $(\alpha,\beta) \in \Lambda$, with $(\alpha,\beta) < (\mu,\omega)$, the set $\{(\alpha',\beta') \in \Lambda : (\alpha',\beta') \le (\alpha,\beta)\}$ has cardinality less than or equal to

$$\text{card } \alpha \cdot \sup\big\{\text{card } \mu_{\alpha'} :\ \alpha' \le \alpha\big\} \le \text{card } \alpha \cdot \text{card } \alpha = \text{card } \alpha < \text{card } \mu.$$

It then easily follows that the whole set Λ can be enumerated by the interval $[\omega,\mu]$. Now put

$$Q_{(\alpha,\beta)} = Q_\beta^\alpha \circ (P_{\alpha+1} - P_\alpha) + P_\alpha, \qquad \omega \le \beta < \mu_\alpha, \qquad \omega \le \alpha < \mu,$$

and let $Q_{(\mu,\omega)}$ be the identity mapping on V.

It remains to verify the properties (i)–(iv) from Definition 6.2.6. Properties (i) and (ii) are obvious. Property (iii) holds true because $P_{\alpha+1} - P_\alpha$ and $P_{\alpha'+1} - P_{\alpha'}$ are "orthogonal" if $\alpha \ne \alpha'$ and because Q_β^α "operate" in the space $(P_{\alpha+1} - P_\alpha)V$. As regards to (iv), we have, according to Proposition 6.2.1(iii), that, for every $v \in V$,

$$v \in \overline{\text{sp}}\Big\{ \bigcup (P_{\alpha+1} - P_\alpha)v :\ \omega \le \alpha < \mu \Big\}$$

$$\subset \overline{\text{sp}}\Big\{ \bigcup \big\{ \bigcup (Q_{\beta+1}^\alpha - Q_\beta^\alpha) \circ (P_{\alpha+1} - P_\alpha)v :\ \omega \le \beta < \mu_\alpha \big\} :\ \omega \le \alpha < \mu \Big\}$$

$$= \overline{\text{sp}}\Big\{ \bigcup (Q_{(\alpha,\beta+1)} - Q_{(\alpha,\beta)})v :\ \omega \le \beta < \mu_\alpha,\ \omega \le \alpha < \mu \Big\}.$$

$\square$

6.3. NOTES AND REMARKS

The first P.R.I.'s were constructed by Lindenstrauss [L1, L2]; see also [AL]. Similar ideas were then used by Tacon [T] who found a P.R.I. in certain dual spaces. An elegant topological method for obtaining a P.R.I. is due to Gul'ko [Gu, NW]. Some of his ideas, translated into the language of Banach spaces, can be found in Valdivia [Va1]. An involved form of a projective generator can be traced back to John and Zizler [JZ1, JZ2]. Projective generators in a more explicit form, in some dual spaces, can be found in the papers [F2, F3, FG]. Finally, Orihuela and Valdivia introduced in [OV] the concept of projective generator in order to unify the constructions from [FG] and [Va1]. Our definition is a slight variant of that from [OV]. Projective generators in some concrete classes of spaces will be constructed in the next two chapters.

The S.P.R.I. can be found, implicitly, in [Tr1] and, explicitly, in [Z2]. Our explanation in Section 6.1 is an elaboration of the papers mentioned above. Lemma 6.1.2 is due to Orihuela [O].

Results similar to those presented in Section 6.2 are almost as old as the P.R.I.; see [AL, T, JZ2, JZ3].

Recent progress in Markuševič bases may be found in [VWZ].

Not every Banach space has a P.R.I. Pličko showed that *the dual of the James tree space* [J] *contains a subspace without a P.R.I.* [Pl].

For dual spaces we have the following result due to Heinrich and Mankiewicz: *The dual of every nonseparable Banach space has uncountably many nontrivial linear norm 1 projections* [HM]. In the proof, they used model theory. For an elementary proof of this fact, see the paper by Sims and Yost [SY]. However, there is no guarantee that such projections would form a P.R.I. or would be, in general, weak*-to-weak* continuous.

As regards to the existence of at least one projection, Gowers and Maurey independently *constructed a separable Banach space such that none of its subspaces admits a nontrivial linear bounded projection;* this fact was observed by W. B. Johnson [GM]. This space gave a stronger (negative) answer to an old question of Lindenstrauss: *Can every Banach space be decomposed as a topological direct sum of two infinitely dimensional subspaces?* The analogous question for nonseparable Banach spaces seems to be open. For a survey on this and other related questions, like the unconditional basic sequence problem and Banach's hyperplane problem (both of which have recently been answered), see [Gow2].

Finally, let us mention a further important consequence of the existence of a P.R.I. *If a Banach space V has a P.R.I.* $\{P_\alpha : \omega \le \alpha \le \mu\}$ *and, if for each $\alpha \in [\omega, \mu)$, the subspace $(P_{\alpha+1} - P_\alpha)V$ admits an equivalent LUR norm, then there exists an equivalent LUR norm on the whole space V.* This result is Zizler's variant [Z1] of Troyanski's famous renorming technique [Tr1]; see also [DGZ2, Proposition VII.1.5].

Chapter Seven

Vašák Spaces
and Gul'ko Compacta

Vašák spaces are a generalization of WCG spaces. Their topological counterpart, Gul'ko compacta, extend the class of Eberlein compacta. In the first section, we collect some basic properties of $\mathcal{K}$-countably determined spaces, Vašák spaces, and Gul'ko compacta. The aim of the second section is to show that the dual of a Vašák space has a covering property, which implies fragmentability; this is done with the help of P.R.I. Thus, Vašák spaces belong to class $\widetilde{\mathcal{F}}$, which is included in $\widetilde{\mathcal{S}}$. In particular, Vašák spaces are weak Asplund. In the third section, we show, in the way of a counterexample, that the weak Asplundness goes beyond the Vašák spaces.

7.1. BASICS ABOUT $\mathcal{K}$-COUNTABLE DETERMINACY

We put $S = \{\emptyset\} \cup \mathbb{N} \cup \mathbb{N}^2 \cup \cdots$, and we endow $\mathbb{N}^{\mathbb{N}}$ with the product topology. Given $\sigma \in \mathbb{N}^{\mathbb{N}}$ and $n \in \mathbb{N}$, $\sigma|n$ denotes the finite sequence $\{\sigma(1), \ldots, \sigma(n)\}$. For $s \in S$ the symbol $|s|$ means the cardinality of s. If $\sigma \in \mathbb{N}^{\mathbb{N}}$ and $s \in S$, then $\sigma \succ s$ means that $\sigma|n = s$, where $n = |s|$. Let X, Y be two topological spaces and $\varphi : X \to 2^Y$ be a multivalued mapping. φ is called *upper semicontinuous* if for any $x \in X$ and any open subset W in Y, with $W \supset \varphi(x)$, there is a neighborhood U of x such that $\varphi(U) \subset W$. If in addition to being upper semicontinuous $\varphi(x)$ is a nonempty compact set for every $x \in X$, φ is called *usco*. The letter p denotes the topology of pointwise convergence on $C(K)$.

Proposition 7.1.1.

Let $Y \neq \emptyset$ be a subspace of a compact space K. Then the following assertions are equivalent:

(i) There exist closed sets $A_n \subset K$, $n \in \mathbb{IN}$, with the property that for every $y \in Y$ and for every $k \in K \backslash Y$ there exists $n \in \mathbb{IN}$ such that $y \in A_n$ and $k \notin A_n$;

(ii) There exist closed sets $B_s \subset K$, $s \in S$, and $\Sigma' \subset \mathbb{IN}^{\mathbb{IN}}$ such that

$$Y = \bigcup_{\sigma \in \Sigma'} \bigcap_{n=1}^{\infty} B_{\sigma|n};$$

(iii) There exist a subset $\Sigma' \subset \mathbb{IN}^{\mathbb{IN}}$ and a usco multivalued mapping φ from Σ' onto Y;

(iv) There exist a subset $\Sigma' \subset \mathbb{IN}^{\mathbb{IN}}$, a closed set $C \subset \Sigma' \times K$, and a continuous (single-valued) mapping f from C onto Y.

Proof. (i)$\Rightarrow$(ii). For $s = \{n_1, n_2, \ldots, n_m\} \in S$ we put $B_s = A_{n_1} \cap A_{n_2} \cap \cdots \cap A_{n_m}$. Then we define

$$\Sigma' = \left\{ \sigma \in \mathbb{IN}^{\mathbb{IN}} : \emptyset \neq \bigcap_{n=1}^{\infty} B_{\sigma|n} \subset Y \right\}.$$

(ii)$\Rightarrow$(iii). Pick a $y \in Y$ and replace each B_s by $B_s \cup \{y\}$. For $\sigma \in \Sigma'$ we put $\varphi(\sigma) = \bigcap_{n=1}^{\infty} B_{\sigma|n}$. Then, surely, $\varphi(\sigma)$ is a nonempty compact subset of Y. The multivalued mapping φ is also surjective. Now consider $\sigma \in \Sigma'$ and an open set $\varphi(\sigma) \subset U \subset Y$. This means that $\bigcap_{n=1}^{\infty} B_{\sigma|n} \subset U$. Assume that for every $m = 1, 2, \ldots$ there is $y_m \in \bigcap_{n=1}^{m} B_{\sigma|n} \backslash U$. As K is compact, $\{y_m\}$ has, in K, a cluster point, k, say. Then $k \in \bigcap_{n=1}^{\infty} B_{\sigma|n} \backslash U$ ($= \varphi(\sigma) \backslash U = \emptyset$), a contradiction. Hence there is $m \in \mathbb{IN}$ such that $\bigcap_{n=1}^{m} B_{\sigma|n} \subset U$. From this we can then conclude that $\varphi(\Omega) \subset U$, where $\Omega = \{\tau \in \Sigma' : \tau \succ \sigma|m\}$, and this set is open. Hence φ is upper semicontinuous at σ.

(iii)$\Rightarrow$(iv). Let C be (the graph of) φ, that is, $C = \bigcup \{\{\sigma\} \times \varphi(\sigma) : \sigma \in \Sigma'\}$. Thus C projects, along Σ', onto Y. That C is closed follows from Lemma 3.1.1.

(iv)$\Rightarrow$(iii). By diminishing Σ', if necessary, we may and do assume that the projection of C along K is all of Σ'. For $\sigma \in \Sigma'$ we put $\varphi(\sigma) = f(({\{\sigma\} \times K}) \cap C)$; then surely $\varphi(\sigma) \neq \emptyset$. As f is continuous, each $\varphi(\sigma)$ is compact. Further, φ maps Σ' onto Y. Now, consider $\sigma \in \Sigma'$ and an open set $\varphi(\sigma) \subset U \subset Y$ and assume that there are $\sigma_n \in \Sigma'$, with $\sigma_n \to \sigma$, and $y_n \in \varphi(\sigma_n) \backslash U$. (We may work with sequences only because Σ' is metrizable.) Find $k_n \in K$ such that $y_n = f(\sigma_n, k_n)$, $n \in \mathbb{IN}$. Let $\{k_{n_\tau}\}$ be a subnet converging to some $k \in K$. It follows that $(\sigma, k) \in C$ and $f(\sigma, k) \in \varphi(\sigma) \subset U$. However, $f(\sigma, k) = \lim_\tau f(\sigma_{n_\tau}, k_{n_\tau}) \notin U$, a contradiction. Therefore φ is upper semicontinuous at σ.

(iii)$\Rightarrow$(ii). For $s \in S$ let B_s be the closure in K of the set $\bigcup \{\varphi(\sigma) : \sigma \in \Sigma', \sigma \succ s\}$. Fix $y \in Y$. Since φ is surjective, there is a $\sigma \in \Sigma'$ such that $y \in \varphi(\sigma)$. Thus $y \in \varphi(\sigma) \subset B_{\sigma|n}$ for each $n \in \mathbb{IN}$. This proves the inclusion "$\subset$".

It remains to show that $\bigcap_{n=1}^{\infty} B_{\sigma|n}$ lies in Y for every $\sigma \in \Sigma'$. To do this, we shall verify that even

$$\bigcap_{n=1}^{\infty} B_{\sigma|n} = \varphi(\sigma).$$

So take $\sigma \in \Sigma'$ and $k \in K \backslash \varphi(\sigma)$. Then $\varphi(\sigma)$ lies in the open set $Y \backslash \{k\}$. Let U be an open set in K such that $\varphi(\sigma) \subset U \subset \bar{U} \subset K \backslash \{k\}$. The upper semicontinuity of φ now yields an $m \in$ IN such that $\varphi(\tau) \subset U$ whenever $\tau \in \Sigma'$ and $\tau \succ \sigma|m$. Hence $B_{\sigma|m} \subset \bar{U}$ and so $k \notin B_{\sigma|m} \left(\supset \bigcap_{n=1}^{\infty} B_{\sigma|n} \right)$.

(ii)$\Rightarrow$(i). The set S is countable so that we can write $S = \{s_n\}$. Therefore it is enough to put $A_n = B_{s_n}$, $n \in$ IN. $\qquad\square$

Definition 7.1.2. A topological space Y is called $\mathcal{K}$-*countably determined* ($\mathcal{K}$-c.d.) if there exists a compact space K such that Y is a subspace of K (i.e., Y is compeletely regular) and the assertions or Proposition 7.1.1 are satisfied.

It should be noted that K in the above definition may be replaced by any other compact superspace of Y; see (iii) in Proposition 7.1.1.

Theorem 7.1.3.

(i) *If a subspace of a compact space is a continuous image of a* $\mathcal{K}$-*c.d. space, then it is* $\mathcal{K}$-*c.d., too.*

(ii) *A closed subspace of a* $\mathcal{K}$-*c.d. space is* $\mathcal{K}$-*c.d.*

(iii) *If* Y_n, $n = 1, 2, \ldots$, *are* $\mathcal{K}$-*c.d., then so is* $\prod_{n=1}^{\infty} Y_n$.

(iv) *If* $Y = \bigcup_{n=1}^{\infty} Y_n$ *is a subspace of a compact space and each subspace* Y_n *is* $\mathcal{K}$-*c.d., then* Y *is* $\mathcal{K}$-*c.d.*

(v) *If* Y_n, $n = 1, 2, \ldots$, *are* $\mathcal{K}$-*c.d. and lie in a compact space, then* $\bigcap_{n=1}^{\infty} Y_n$ *is* $\mathcal{K}$-*c.d.*

Proof. (i) This holds because the composition of a usco mapping and a continuous mapping is again a usco mapping.

(ii) Let Z be a closed subspace of a $\mathcal{K}$-c.d. space Y and pick $z \in Z$; then Z is also completely regular. Let φ and Σ' correspond to Y as in Definition 7.1.1. Put $\widetilde{\varphi}(\sigma) = (\varphi(\sigma) \cap Z) \cup \{z\}$, $\sigma \in \Sigma'$. Then $\widetilde{\varphi}$ is, by Lemma 3.1.1, a usco mapping from Σ' onto Z.

(iii) Assume that each Y_n is $\mathcal{K}$-c.d. and is a subset of a compact space K_n. For $n = 1, 2, \ldots$ let $\Sigma_n \subset$ IN^{IN} and $\varphi_n : \Sigma_n \to Y_n$ be usco mappings such that $\varphi_n(\Sigma_n) = Y_n$. We remark that the space $(\text{IN}^{\text{IN}})^{\text{IN}}$ is homeomorphic with IN^{IN} under the mapping ψ defined by

$\psi(\{\sigma_1, \sigma_2, \ldots\})$

$\qquad = \{\sigma_1(1), \sigma_2(1), \sigma_1(2), \sigma_3(1), \sigma_2(2), \sigma_1(3), \ldots\}, \quad \{\sigma_1, \sigma_2, \ldots\} \in (\text{IN}^{\text{IN}})^{\text{IN}}.$

Put now $\Sigma' = \psi\left(\prod_{n=1}^{\infty} \Sigma_n\right)$ and define $\varphi : \Sigma' \to \prod_{n=1}^{\infty} Y_n$ by

$$\varphi(\sigma) = \prod_{n=1}^{\infty} \varphi_n\left(\psi^{-1}(\sigma)_n\right)$$

where $\psi^{-1}(\sigma)_n$ means the n-th coordinate of $\psi^{-1}(\sigma)$. Using Lemma 3.1.1, it is a routine matter to verify that φ is both usco and surjective. This, together with the fact that $\prod_{n=1}^{\infty} Y_n$ lies in the compact space $\prod_{n=1}^{\infty} K_n$ completes the proof of (iii).

Let Y be as in (iv). Pick $y \in Y$ and for each $n \in \mathbb{N}$ replace Y_n by $Y_n \cup \{y\}$. Clearly, the new Y_n will also be $\mathcal{K}$-c.d. Put then

$$Z = \left\{ \{y_n\} \in \prod_{n=1}^{\infty} Y_n : \ y_n \neq y \quad \text{for at most one } \ n \in \mathbb{N} \right\}.$$

Then Z is closed in $\prod_{n=1}^{\infty} Y_n$. Hence, by (iii) and (ii), Z is $\mathcal{K}$-c.d. We define $\psi : Z \to Y$ by $\psi(\{y_n\}) = y_m$, where $m \in \mathbb{N}$ is such that $y_n = y$ for all $n \in \mathbb{N}\backslash\{m\}$. It is easy to check that ψ is continuous and that $\psi(Z) = \bigcup_{n=1}^{\infty} Y_n \ (= Y)$. Hence, by (i), Y is $\mathcal{K}$-c.d.

(v) If each Y_n is $\mathcal{K}$-c.d., then so is their product, Y, say. Put $Z = \{\{y_n\} \in Y : y_1 = y_2 = \dots\}$; this is a closed subset of Y. Hence Z is $\mathcal{K}$-c.d. Now the assignment $\{y_n\} \mapsto y_1$ maps Z continuously onto $\bigcap_{n=1}^{\infty} Y_n$. Thus $\bigcap_{n=1}^{\infty} Y_n$ is $\mathcal{K}$-c.d. $\qquad\square$

Theorem 7.1.4.

Every $\mathcal{K}$-c.d. space is Lindelöf, that is, every cover of it consisting of open sets admits a countable subcover.

Proof. Let Y be a $\mathcal{K}$-c.d. space; so $Y = \varphi(\Sigma')$ for some $\Sigma' \subset \mathbb{N}^{\mathbb{N}}$ and some usco mapping φ. Let $\{U_\gamma : \ \gamma \in \Gamma\}$ be an open cover of Y. Take any $\sigma \in \Sigma'$. As $\varphi(\sigma)$ is compact, there is a finite set $F_\sigma \subset \Gamma$ such that $\varphi(\sigma) \subset \bigcup\{U_\gamma : \ \gamma \in F_\sigma\}$. Since φ is upper semicontinuous, there is an open neighborhood W_σ of σ such that $\varphi(W_\sigma) \subset \bigcup\{U_\gamma : \ \gamma \in F_\sigma\}$. Now, since Σ' is a separable metrizable space, it is Lindelöf [En, Theorem 3.8.1]. Hence there is a sequence $\{\sigma_n\} \subset \Sigma'$ such that $\Sigma' = \bigcup_{n=1}^{\infty} W_{\sigma_n}$. Then

$$Y = \varphi(\Sigma') = \bigcup_{n=1}^{\infty} \varphi(W_{\sigma_n}) \subset \bigcup\{U_\gamma : \ \gamma \in F_{\sigma_n}, \ n = 1, 2, \dots\}$$

and the family on the right-hand side is countable. Hence Y is Lindelöf. $\qquad\square$

Definition 7.1.5. A Banach space V is called *Vašák* or *weakly countably determined*, or just *WCD*, if (V, w) is $\mathcal{K}$-c.d.

Here it should be noted that any Banach space endowed with the weak topology is completely regular; see the argument below Definition 3.2.1.

Proposition 7.1.6.

Subspaces of WCG spaces are Vašák and hence Lindelöf in the weak topology.

Proof. Let Z be a subspace of a WCG space V. Then there is a weakly compact set K in V such that $\overline{\text{sp}}K = V$. We denote by $\mathbf{A}$ the set of all sequences $\{\alpha_n\}$ of rational numbers such that $\alpha_m = 0$ for all but finitely many $m \in \text{IN}$. The set $\mathbf{A}$ is countable; enumerate it as $\mathbf{A} = \{a_n\}$. For $n \in \text{IN}$ put $K_n = \{\alpha_1 k_1 + \alpha_2 k_2 + \cdots : k_i \in K\} \cup \{0\}$ where $\{\alpha_i\} = a_n$; we observe that the sets K_n are weakly compact. Finally, for $\sigma \in \text{IN}^{\text{IN}}$ we put

$$\varphi(\sigma) = \bigcap_{n=1}^{\infty} \left(K_{\sigma(n)} + \frac{1}{n} B_{V^{**}} \right).$$

The sets $\varphi(\sigma)$ are nonempty and compact in (V^{**}, w^*) and moreover lie in V. Therefore they are compact in (V, w). Now take any $v \in V$. Since $\overline{\text{sp}}\,K = V$, for each $n \in \text{IN}$ there is $i_n \in \text{IN}$ such that $v \in K_{i_n} + (1/n)B_V$. Hence, putting $\sigma = \{i_1, i_2, \ldots\}$, we have $v \in \varphi(\sigma)$. Therefore $\varphi(\text{IN}^{\text{IN}}) = V$. Finally, it is an immediate consequence of the definition of φ that it is upper continuous as a mapping from IN^{IN} to (V^{**}, w^*) and so as a mapping from IN^{IN} to (V, w). Thus V is Vašák and so Z is a Vašák space by Proposition 7.1.3(ii).

The weak Lindelöfness of V immediately follows from Definition 7.1.5 and Theorem 7.1.4. □

Definition 7.1.7. A compact space K is called a *Gul'ko* compact if the space $C(K)$ endowed with the pointwise topology p is $\mathcal{K}$-c.d.

We note that $(C(K), p)$ is also completely regular. This can be shown by a similar argument as proving that (V, w) is completely regular.

That Eberlein compacta are Gul'ko compacta will be proved in the next section.

Theorem 7.1.8.

For a compact space K the following assertions are equivalent:
  (i) K is a Gul'ko compact;
  (ii) $C(K)$ is a Vašák space; and
  (iii) There exists a $\mathcal{K}$-c.d. subspace Y of $(C(K), p)$ which separates the points of K, that is, if $k, k' \in K$ are different, there is $f \in Y$ such that $f(k) \neq f(k')$.

Proof. The implication (i)⇒(iii) is trivial. The implication (ii)⇒(i) holds because the pointwise topology is weaker than the weak topology.

(i)⇒(ii). Assume that $(C(K), p)$ is $\mathcal{K}$-c.d. Then $(B_{C(K)}, p)$ is $\mathcal{K}$-c.d. by Theorem 7.1.3(ii). We shall need the claim: *A subset of $B_{C(K)}$ is weakly compact if (and only if) it is compact in the topology* p. Assume this has already been proved. Let φ be a usco mapping from some $\Sigma' \subset \mathrm{IN}^{\mathrm{IN}}$ onto $(B_{C(K)}, p)$. Then φ, considered as a mapping to $(B_{C(K)}, w)$, is also usco. In fact, from the claim we know that $\varphi(\sigma)$ is weakly compact for each $\sigma \in \Sigma'$. Further consider $\sigma \in \Sigma'$, an open set $U \subset (B_{C(K)}, w)$ with $U \supset \varphi(\sigma)$, and assume that there are $\sigma_n \in \Sigma'$, with $\sigma_n \to \sigma$, and such that $\varphi(\sigma_n) \backslash U \neq \emptyset$ for all $n = 1, 2, \ldots$. (We may restrict ourselves to working with sequences because Σ' is metrizable.) Since φ is p-usco, an easy excercise reveals that $\varphi(\{\sigma, \sigma_n, \sigma_{n+1}, \ldots\})$, $n \in \mathrm{IN}$, are p-compact sets lying in $B_{C(K)}$; see the proof of Theorem 4.1.2. By the claim, these sets are weakly compact; thus $\varphi(\{\sigma, \sigma_n, \sigma_{n+1}, \ldots\}) \backslash U$ are weakly compact. Hence, there is f in $\bigcap_{n=1}^{\infty} \varphi(\{\sigma, \sigma_n, \sigma_{n+1}, \ldots\}) \backslash U$. Then $f \notin \varphi(\sigma)$ and so $\varphi(\sigma) \subset B_{C(K)} \backslash \{f\}$, the last set being p-open in $B_{C(K)}$. Now the p-upper semicontinuity of φ guarantees that $\varphi(\sigma_n) \subset B_{C(K)} \backslash \{f\}$ for all large $n \in \mathrm{IN}$, a contradiction.

We have proved that φ is a usco mapping from Σ' onto $(B_{(C(K)}, w)$. This means that $(B_{C(K)}, w)$ is $\mathcal{K}$-c.d. Theorem 7.1.3(iv) finally gives us that $(C(K), w)$ is $\mathcal{K}$-c.d., that is, $C(K)$ is Vašák and so (ii) is proved.

It remains to prove the claim. Let Y be a p-compact set in $B_{C(K)}$. According to the Eberlein–Šmulyan theorem, it is enough to show that every sequence in Y has a weakly convergent subsequence. So let $\{f_n\}$ be a sequence in Y. By diminishing Y, if necessary, we may and do assume that $\{f_n\}$ is p-dense in Y. We define $\phi : K \to [-1, 1]^{\mathrm{IN}}$ by

$$\phi(k)(n) = f_n(k), \qquad n \in \mathrm{IN}, \qquad k \in K$$

and put $L = \phi(K)$. Further let $\psi : Y \to \mathrm{IR}^L$ be defined by

$$\psi(y)(l) = y(k), \qquad y \in Y, \qquad l \in L, \qquad \text{and} \qquad k \in \phi^{-1}(l).$$

In the proof of Theorem 1.1.3, it is shown that such a ψ is well defined and that it maps Y into $C(L)$. Since L is separable, there is a sequence $\{k_i\} \subset K$ such that $\{\phi(k_i)\}$ is dense in L. Consider now the pseudometric ρ on Y defined by

$$\rho(y_1, y_2) = \sum_{i=1}^{\infty} |y_1(k_i) - y_2(k_i)| 2^{-i}, \qquad y_1, y_2 \in Y.$$

Then ρ is in fact a metric since $\rho(y_1, y_2) = 0$ implies $y_1(k_i) = y_2(k_i)$ for all $i = 1, 2, \ldots$. This means that $\psi(y_1)(\phi(k_i)) = \psi(y_2)(\phi(k_i))$ for all i. Hence $\psi(y_1)(l) = \psi(y_2)(l)$ for all $l \in L$ and so $y_1(k) = y_2(k)$ for all $k \in K$. Finally, since Y is in $B_{C(K)}$, the identity mapping $(Y, p) \to (Y, \rho)$ is continuous. It follows (Y, p) is a metrizable compact. Therefore $\{f_n\}$ has a pointwise convergent subsequence. Then F. Riesz's theorem and Lebesgue's dominated convergence theorem say that this subsequence converges weakly.

(iii)$\Rightarrow$(i). Assume that (iii) is satisfied. We shall use Theorem 7.1.3 many times. If $Z \subset (C(K), p)$ is $\mathcal{K}$-c.d., then so are $\mathbb{R}Z$, $Z \times Z$, $Z \pm Z :=$ $\{f \pm g : f, g \in Z\}$ and $Z \cdot Z := \{f \cdot g : f, g \in Z\}$, where $(f \cdot g)(k) = f(k)g(k)$, $k \in K$. It follows that the smallest algebra W containing Y and all the constant functions is also $\mathcal{K}$-c.d. Now, we know from the Stone–Weierstrass theorem that W is norm-dense in $C(K)$. Put $B = \{f \in W : \|f\| \le 1\}$; this is a closed set in (W, p) and so $\mathcal{K}$-c.d. Thus, $(W, p) \times (B, p)^{\mathbb{N}}$ is $\mathcal{K}$-c.d. Finally, (i) holds, that is, $(C(K), p)$ is $\mathcal{K}$-c.d., since there is a continuous mapping Ψ from $(W, p) \times (B, p)^{\mathbb{N}}$ onto $(C(K), p)$ defined by

$$\Psi(f, \{f_n\})(k) = f(k) + \sum_{n=1}^{\infty} 2^{-n} f_n(k), \qquad f \in W, \qquad \{f_n\} \in B^{\mathbb{N}}, \qquad k \in K.$$

Let us show that Ψ is in fact surjective. Take any $g \in C(K)$. Since W is dense in $C(K)$, there are, subsequently, $f \in W$ with $\|g - f\| < 2^{-2}$, $f_1 \in W$ with $\|g - f - f_1\| < 2^{-3}, \ldots, f_n \in W$ with $\|g - f - f_1 - \cdots - f_n\| < 2^{-n-2}, \ldots$. Then we easily get that $\|f_n\| < 2^{-n}$, $n = 1, 2, \ldots$. Hence $g = f + \sum_{n=1}^{\infty} f_n$ in norm, $2^n f_n \in B$, and so $g = \Psi(f, \{2^n f_n\})$. $\qquad\qquad\square$

Theorem 7.1.9.

For a Banach space V the following assertions are equivalent:
(i) V is a Vašák space;
(ii) (B_{V^*}, w^*) is a Gul'ko compact; and
(iii) There exists a set $Y \subset V$ such that (Y, w) is $\mathcal{K}$-c.d. and $\overline{\mathrm{sp}}\, Y = V$.

Proof. That (i) implies (iii) is trivial. Let (iii) be satisfied. Let κ denote the canonical embedding of (V, w) into $(C(B_{V^*}, w^*), p)$, that is, $\kappa(v)(v^*) = \langle v^*, v \rangle$, $v \in V$, $v^* \in V^*$. Then $(\kappa(Y), p)$ is $\mathcal{K}$-c.d. by Theorem 7.1.3(i). Further $\kappa(Y)$ separates the points of B_{V^*}. Hence, by Theorem 7.1.8, (B_{V^*}, w^*) is a Gul'ko compact and (ii) is proved. Finally, let (ii) be satisfied. Then, by Theorem 7.1.8, the space $C(B_{V^*}, w^*)$ is Vašák, that is, $\mathcal{K}$-c.d. in its weak topology. Now, κ is a linear isometry into, so $\kappa(V)$ is weakly closed in $C(B_{V^*}, w^*)$. Theorem 7.1.3(ii) then guarantees that $(\kappa(V), w)$ is $\mathcal{K}$-c.d., (V, w) is $\mathcal{K}$-c.d., and therefore V is Vašák. $\qquad\qquad\square$

Now permanence properties for Gul'ko compacta and Vašák spaces follow.

Theorem 7.1.10.

The class of Gul'ko compacta is stable when going to
(i) continuous images,
(ii) closed subspaces,

(iii) countable products, and

(iv) finite unions.

(v) If K is a Gul'ko compact, then so is $(B_{C(K)^*}, w^*)$.

Proof. (i) Let K be a Gul'ko compact and consider a continuous surjective mapping $\psi : K \to L$. Then the mapping $f \mapsto f \circ \psi$ sends $C(L)$ isometrically onto a subspace Y of $C(K)$. It is clear that this mapping is a homeomorphism from $(C(L), p)$ onto (Y, p). We shall show that Y is closed in $(C(K), p)$. Let g be in the p-closure of Y. Define $f : L \to \mathbb{R}$ by $f(\psi(k)) = g(k)$, $k \in K$. It is easy to check that f is well defined. It remains to show that f is continuous. So, consider any closed set $C \subset \mathbb{R}$. Then $K \cap g^{-1}(C)$ is a closed set and so is $\psi(K \cap g^{-1}(C))$. Now, it remains to make an elementary observation that the last set is equal to $f^{-1}(C)$. Therefore $f \in C(L)$ and the p-closedness of Y is proved. Now Theorem 7.1.3(ii) applies. Another proof can be given by putting together Theorems 7.1.8 and 7.1.3(ii).

(ii) Let L be a closed subspace of a Gul'ko compact K and consider the restriction mapping Q which assigns $f_{|L}$ to each $f \in C(K)$. Then Q maps $(C(K), p)$ continuously to $(C(L), p)$. It is in fact a surjection according to the Tietze–Urysohn theorem. Hence, by Theorem 7.1.3(i), $(C(L), p)$ is $\mathcal{K}$-c.d., that is, L is a Gul'ko compact.

(iii) Let K_n, $n = 1, 2, \ldots$, be Gul'ko compacta and denote $K = \prod_{n=1}^{\infty} K_n$. For $n = 1, 2, \ldots$ we define $\psi_n : \prod_{i=1}^{\infty} (C(K_i), p) \to (C(K), p)$ by

$$\psi_n(\{f_i\})(\{k_i\}) = f_n(k_n), \qquad \{f_i\} \in \prod_{i=1}^{\infty} (C(K_i), p), \qquad \{k_i\} \in K.$$

Clearly, each ψ_n is well defined and continuous. Let Y denote the union of the ranges of ψ_n, $n = 1, 2, \ldots$. Then (Y, p) is $\mathcal{K}$-c.d. by Theorem 1.7.3. Since Y separates the points of K, Theorem 7.1.8 guarantees that K is a Gul'ko compact.

(iv) Clearly, it is enough to deal with the union of two compacta. So consider two Gul'ko compacta K and L which are subspaces of a compact space H and assume that $K \cup L = H$. Let Z be the set of all couples $(f, g) \in C(K) \times C(L)$ such that $f(h) = g(h)$ whenever $h \in K \cap L$. It is straightforward to check that the set Z is closed in $(C(K), p) \times (C(L), p)$. Define $\psi : Z \to (C(H), p)$ as $\psi(f, g)(h) = f(h)$ if $h \in K$ and $\psi(f, g)(h) = g(h)$ if $h \in H \backslash K$. The mapping ψ is well defined since $\psi(f, g)^{-1}(C) = f^{-1}(C) \cup g^{-1}(C)$ for every closed set $C \subset \mathbb{R}$. Also, ψ is obviously surjective and continuous. Now it remains to apply Theorem 7.1.3.

(v) This follows by putting together Theorems 7.1.8 and 7.1.9. $\qquad \square$

Theorem 7.1.11.

(i) If $T : V \to Y$ is a continuous linear mapping from a Vašák space V into a Banach space Y, with $\overline{TV} = Y$, then Y is Vašák; in particular, quotients of a Vašák space are Vašák.

(ii) Subspaces of a Vašák space are Vašák.

(iii) If V_n, $n = 1, 2, \ldots$, are Vašák spaces, then so are $\left(\sum_{n=1}^{\infty} V_n \right)_{c_0}$ and $\left(\sum_{n=1}^{\infty} V_n \right)_{\ell_p}$, $1 \le p < +\infty$.

(iv) If Vašák spaces V_n, $n = 1, 2, \ldots$, are subspaces of a Banach space V and $\bigcup_{n=1}^{\infty} V_n$ is linearly dense in V, then V is Vašák.

Proof. Under the assumptions of (i), Theorem 7.1.3(i) says that (TV, w) is $\mathcal{K}$-c.d. Theorem 7.1.9 then guarantees that V is Vašák.

(ii) This is a special case of Theorem 7.1.3(ii).

(iii) Assume that the spaces V_n, $n = 1, 2, \ldots$, are Vašák. Denote by ψ the assignment $\{v_n\} \mapsto \{2^{-n} v_n\}$. This maps $\prod_{n=1}^{\infty} (B_{V_n}, w)$ continuously into both $\left(\left(\sum_{n=1}^{\infty} V_n \right)_{c_0}, w \right)$ and $\left(\left(\sum_{n=1}^{\infty} V_n \right)_{\ell_p}, w \right)$ and the range of ψ is linearly dense in the last two spaces. Now it remains to apply Theorems 7.1.3 and 7.1.9.

(iv) This follows from Theorem 7.1.3(iv) and Theorem 7.1.9. □

7.2. DEEPER FACTS ON VAŠÁK SPACES AND GUL'KO COMPACTA

Proposition 7.2.1.

Every Vašák space admits a projectional generator; see Definition 6.1.6.

Proof. Let V be a Vašák space. Hence there exist a subset $\Sigma' \subset \mathbb{N}^{\mathbb{N}}$ and a usco mapping φ from Σ' onto (B_V, w). Put

$$Y_s = \bigcup \{\varphi(\sigma) : \sigma \in \Sigma', \ \sigma \succ s\}, \qquad s \in S.$$

(Recall that $S = \emptyset \cup \mathbb{N} \cup \mathbb{N}^2 \cup \cdots$.) For every $\xi \in V^*$ and every $s \in S$ we find a countable set $\Phi_s(\xi) \subset Y_s$ such that

$$\inf \langle \xi, Y_s \rangle = \inf \langle \xi, \Phi_s(\xi) \rangle$$

and put then

$$\Phi(\xi) = \bigcup \{\Phi_s(\xi) : s \in S\}, \qquad \xi \in V^*.$$

Obviously, $\Phi : V^* \to 2^V$ is a countable valued mapping. We shall show that (V^*, Φ) is a projectional generator on V. To this end, let $\emptyset \ne B \subset V^*$ be any subset with $\overline{B}$ linear. Take ξ in $\Phi(B)^{\perp} \cap \overline{B}^*$ and assume, for the purpose of obtaining a contradiction, that $\xi \neq 0$. Then there are $v_0 \in B_V$ and $r \in \mathbb{R}$ such that $\langle r\xi, v_0 \rangle = -3$. As φ is surjective, there is $\sigma \in \Sigma'$ such that v_0 belongs to $\varphi(\sigma)$.

We claim that

$$\left(r\xi + \varphi(\sigma)^\circ\right) \cap \overline{B} \neq \emptyset,$$

where $\varphi(\sigma)^\circ = \{v^* \in V^* : \sup |\langle v^*, \varphi(\sigma) \rangle| < 1\}$. (This is a consequence of the Mackey–Arens theorem [RR, Theorem III.7].) Suppose, to the contrary, that the claim is false. Because $\varphi(\sigma)$ is bounded, $r\xi + \varphi(\sigma)^\circ$ has a nonempty interior and we may use the separation theorem. Thus, there is $F \in V^{**}$ such that

$$\inf\langle F, r\xi + \varphi(\sigma)^\circ \rangle > \sup\langle F, \overline{B} \rangle = 0;$$

we used here the linearity of $\overline{B}$. It follows therefore that

$$\sup\langle -F, \varphi(\sigma)^\circ \rangle < \langle F, r\xi \rangle \quad (< +\infty),$$

and so the bipolar theorem [RR, Theorem II.4.] ensures that $-cF$ lies in the weak* closed convex hull of $\varphi(\sigma)$ for some $c > 0$. (We identify V with its canonical image in V^{**}.) Now, since $\varphi(\sigma)$ is weakly compact, the Krein–Šmulyan theorem guarantees that $-cF$, and hence F, lies in V. However, the inequality

$$\langle r\xi, F \rangle \geq \inf\langle r\xi + \varphi(\sigma)^\circ, F \rangle > \sup\langle \overline{B}, F \rangle = \sup\langle \overline{B}^*, F \rangle$$

implies that $\xi \notin \overline{B}^*$, which is impossible.

By the claim just proved, there exists η in B such that

$$\sup |\langle r\xi - \eta, \varphi(\sigma) \rangle| < 1.$$

Hence

$$\langle \eta, v_0 \rangle = \langle r\xi, v_0 \rangle + \langle \eta - r\xi, v_0 \rangle < -3 + 1 = -2.$$

From the proof of (iv)$\Rightarrow$(iii) in Proposition 7.1.1 we know that

$$\varphi(\sigma) = \bigcap_{n=1}^{\infty} \overline{Y_{\sigma|n}}^*.$$

(Recall that $B_V = \varphi(\Sigma')$ lies in the compact $(B_{V^{**}}, w^*)$.) Thus there is $n \in \mathbb{N}$ such that

$$\sup |\langle r\xi - \eta, Y_{\sigma|n} \rangle| < 1.$$

Note that

$$\inf\langle \eta, Y_{\sigma|n} \rangle \leq \langle \eta, v_0 \rangle < -2.$$

Find then $y_0 \in \Phi_{\sigma|n}(\eta)$ $(\subset Y_{\sigma|n} \cap \Phi(\eta))$ with $\langle \eta, y_0 \rangle < -2$. Therefore,

$$\langle r\xi, y_0 \rangle = \langle \eta, y_0 \rangle + \langle r\xi - \eta, y_0 \rangle < -2 + 1 = -1 \neq 0.$$

However, this is impossible since $\xi \in \Phi(B)^{\perp}$ and $y_0 \in \Phi(\eta) \subset \Phi(B)$.   □

Theorem 7.2.2.

Let V be a Vašák space. Then:
   (i) If V is nonseparable, then it admits a P.R.I. as well as a S.P.R.I; see Definitions 6.1.5 and 6.2.6.
  (ii) There exist a set Γ and a bounded linear one-to-one mapping from V into $c_0(\Gamma)$.
 (iii) There exists a Markuševič basis on V; see Definition 6.2.3.

Proof. We shall proceed by transfinite induction over the density, $\mathrm{dens}\, V$, of V. First let V be a separable Banach space. That V then has a Markuševič basis is shown in [LT, Proposition 1.f.3]. Now, let $\{v_n\}$ be a sequence contained and dense in the unit sphere of V. For $n = 1, 2, \ldots$ we find $f_n \in V^*$ such that $\langle f_n, v_n \rangle = 1 = \|f_n\|$. Then the assignment $v \mapsto \{n^{-1}\langle f_n, v \rangle : n \in \mathrm{IN}\}$, $v \in V$, maps V linearly, continuously, and injectively into c_0.

Let an uncountable cardinal $\aleph$ be given and assume that we have proved our theorem for all Vašák spaces with density less than $\aleph$. Now consider a Vašák space V with $\mathrm{dens}\, V = \aleph$. Let $\{P_\alpha : \omega \leq \alpha \leq \mu\}$ be a P.R.I. on V; its existence is ensured by Propositions 7.2.1 and 6.1.7. For every $\omega \leq \alpha \leq \mu$ the subspace $(P_{\alpha+1} - P_\alpha)V$ has density less than $\aleph$ and moreover, by Theorem 7.1.11(ii), is also Vašák.

Now, in order to complete the proof, it is enough to put together the induction assumption with Propositions 6.2.7, 6.2.2, and 6.2.4.   □

From Theorems 1.2.4, 7.1.6, and 7.1.8 we immediately get the next proposition. (Recall that, in the proof of (i)⇒(ii) in Theorem 1.2.4, we needed Theorem 7.2.2(ii).)

Proposition 7.2.3.

Every Eberlein compact is a Gul'ko compact.

For $\emptyset \neq \Sigma' \subset \mathrm{IN}^{\mathrm{IN}}$ and for a set $\Gamma \neq \emptyset$ we define

$$c_1(\Sigma' \times \Gamma) = \{f \in \ell_\infty(\Sigma' \times \Gamma) : f_{|K \times \Gamma} \in c_0(K \times \Gamma) \text{ for every compact } K \subset \Sigma'\}$$

and we endow this space with the supremum norm. Obviously $c_0(\Sigma' \times \Gamma) \subset c_1(\Sigma' \times \Gamma)$. It is easy to check that ℓ_∞ is isometric to a subspace of $c_1(\Sigma' \times \Gamma)$ if and only if Σ' is not compact.

Theorem 7.2.4.

A Banach space V is a Vašák space if and only if there exist a set $\emptyset \neq \Sigma' \subset \mathbb{IN}^{\mathbb{IN}}$, a set $\Gamma \neq \emptyset$, and a bounded, linear, one-to-one, and weak*-to-pointwise-continuous mapping T from V^* into $c_1(\Sigma' \times \Gamma)$.

Proof. Let V be Vašák; therefore we can write $V = \varphi(\Sigma')$, where $\Sigma' \subset \mathbb{IN}^{\mathbb{IN}}$ and φ is a usco mapping from Σ' onto (V, w). By Theorem 7.2.2(i), there is a S.P.R.I. $\{Q_\alpha : \omega \leq \alpha \leq \mu\}$ on V. For each $\alpha \in [\omega, \mu)$ we find sequences $\{v_n^\alpha : n \in \mathbb{IN}\}$, $\|v_n^\alpha\| < 1/n$, which are contained and are linearly dense in $(Q_{\alpha+1} - Q_\alpha)V$. Further, for every $\alpha \in [\omega, \mu)$ and every $n \in \mathbb{IN}$ we take $\sigma_n^\alpha \in \Sigma'$ such that $v_n^\alpha \in \varphi(\sigma_n^\alpha)$. Finally, we define a mapping $T : V^* \to \ell_\infty(\Sigma' \times \mathbb{IN} \times [\omega, \mu))$ by

$$Tv^*(\sigma, n, \alpha) = \begin{cases} \langle v^*, v_n^\alpha \rangle & \text{if } \sigma = \sigma_n^\alpha, \\ 0 & \text{otherwise;} \end{cases} \qquad (\sigma, n, \alpha) \in \Sigma' \times \mathbb{IN} \times [\omega, \mu).$$

Clearly, T is well defined (as $\|v_n^\alpha\| \leq 1/n$), linear, bounded, and weak*-to-pointwise continuous. It is also injective. To see this, take any $0 \neq v^* \in V^*$. From Definition 6.2.6(iv) we know that $\bigcup\{(Q_{\alpha+1} - Q_\alpha)V : \omega \leq \alpha < \mu\}$ is linearly dense in V. So there is $\alpha \in [\omega, \mu)$ such that v^* restricted to $(Q_{\alpha+1} - Q_\alpha)V$ is not identically zero. Hence, there is an $n \in \mathbb{IN}$ such that

$$\langle v^*, v_n^\alpha \rangle \neq 0.$$

This means that $Tv^*(\sigma_n^\alpha, n, \alpha) \neq 0$ and so $Tv^* \neq 0$.

It remains to prove that the range of T lies in $c_1(\Sigma' \times \mathbb{IN} \times [\omega, \mu))$. So take any $v^* \in V^*$ and fix a compact set $K \subset \Sigma'$. We have to show that Tv^* restricted to $K \times \mathbb{IN} \times [\omega, \mu)$ belongs to $c_0(K \times \mathbb{IN} \times [\omega, \mu))$. In order to obtain a contradiction, assume there exist $\epsilon > 0$ and a one-to-one infinite sequence $\{(\sigma_m, n_m, \alpha_m)\}$ in $K \times \mathbb{IN} \times [\omega, \mu)$ such that

$$|Tv^*(\sigma_m, n_m, \alpha_m)| > \epsilon \quad \text{for all} \quad m \in \mathbb{IN}.$$

Thus σ_m must be of the form $\sigma_{n_m}^{\alpha_m}$ and so

$$\left|\langle v^*, v_{n_m}^{\alpha_m} \rangle\right| > \epsilon \quad \text{for all} \quad m \in \mathbb{IN}.$$

Moreover $n_m^{-1}\|v^*\| > \epsilon$ and so the n_m are allowed to occur in a fixed finite set only. It follows that the sequence $\{\alpha_m\}$ contains infinitely many distinct elements. Let us, for simplicity, assume that $\alpha_m \neq \alpha_k$ whenever $m \neq k$. Now $v_{n_m}^{\alpha_m} \in \varphi(\sigma_{n_m}^{\alpha_m}) = \varphi(\sigma_m) \subset \varphi(K)$, which is a weakly compact set owing to the properties of φ; see the proof of Theorem 4.1.2. Let v be a weak cluster point of the sequence $\{v_{n_m}^{\alpha_m}\}$. Since $\alpha_m \neq \alpha_k$ for $m \neq k$, the "orthogonality" of $\{Q_{\alpha+1} - Q_\alpha\}$ implies that $(Q_{\alpha+1} - Q_\alpha)v = 0$ for each $\alpha \in [\omega, \mu)$. Hence $v = 0$.

However, on the other hand, we have

$$|\langle v^*, v\rangle| \geq \liminf_{m\to\infty}|\langle v^*, v^{\alpha_m}_{n_m}\rangle| \geq \epsilon > 0,$$

which is a contradiction. Therefore, Tv^* lies in $c_1(\Sigma' \times \mathbb{N} \times [\omega, \mu))$ and the necessity is proved.

Conversely, assume we have Σ', Γ, and T as asserted. For $(\sigma, \gamma) \in \Sigma' \times \Gamma$, let $\delta_{(\sigma,\gamma)}$ be the element of $c_1(\Sigma' \times \Gamma)^*$ defined by $\langle \delta_{(\sigma,\gamma)}, f\rangle = f(\sigma, \gamma)$, $f \in c_1(\Sigma' \times \Gamma)$. Define

$$Z = \{\delta_{(\sigma,\gamma)} : (\sigma, \gamma) \in \Sigma' \times \Gamma\};$$

then $Z \subset B_{c_1(\Sigma'\times\Gamma)^*}$. We shall show that (Z, w^*) is $\mathcal{K}$-c.d. For $\sigma \in \Sigma'$ put

$$\varphi(\sigma) = \{\delta_{(\sigma,\gamma)} : \gamma \in \Gamma\} \cup \{0\}.$$

The set $\varphi(\sigma)$ is weak* compact since every $f \in c_1(\Sigma' \times \Gamma)$ has the property that $f_{|\{\sigma\}\times\Gamma} \in c_0(\{\sigma\} \times \Gamma)$. Further we check that φ is upper semicontinuous. Assume this is not the case. Then there exist $\sigma \in \Sigma'$, an open set $\varphi(\sigma) \subset U \subset (Z, w^*)$, a sequence $\{\sigma_n\} \subset \Sigma'$, with $\sigma_n \to \sigma$, and a sequence $\{\gamma_n\} \subset \Gamma$ such that $\delta_{(\sigma_n,\gamma_n)} \notin U$ for all $n \in \mathbb{N}$. (Sequences are enough because Σ' is metrizable.) Now, remarking that the set $\{\sigma, \sigma_1, \sigma_2, \ldots\}$ is compact, we get that $\delta_{(\sigma_n,\gamma_n)} \to 0$ weak*, so $0 \notin U$, a contradiction. Therefore (Z, w^*) is $\mathcal{K}$-c.d.

Then $(T^*(Z), w^*)$ is $\mathcal{K}$-c.d. Moreover $T^*\delta_{(\sigma,\gamma)}$ are weak* continuous functionals on V^*. In fact, if $v^*_\tau \to v^*$ weak*, we have

$$\langle T^*\delta_{(\sigma,\gamma)}, v^*_\tau\rangle = \langle \delta_{(\sigma,\gamma)}, Tv^*_\tau\rangle = Tv^*_\tau(\sigma, \gamma) \to Tv^*(\sigma, \gamma) = \langle T^*\delta_{(\sigma,\gamma)}, v^*\rangle$$

since T is weak*-to-pointwise continuous. It then follows, by [DS, Corollary V.3.11], that $T^*\delta_{(\sigma,\gamma)}$ lies in V. (We assume that V is a subspace of V^{**}.) Further we check that $\mathrm{sp}\{T^*\delta_{(\sigma,\gamma)} : (\sigma, \gamma) \in \Sigma' \times \Gamma\}$ is dense in V. In fact, if it is not, then there is $0 \neq v^* \in V^*$ such that for all $(\sigma, \gamma) \in \Sigma' \times \Gamma$ we have $\langle T^*\delta_{(\sigma,\gamma)}, v^*\rangle = 0$, that is, $Tv^*(\sigma, \gamma) = 0$. However, as T is injective, $v^* = 0$, which is a contradiction. Thus, $(T^*(Z), w)$ being $\mathcal{K}$-c.d., Theorem 7.1.9 guarantees that V is a Vašák space. $\qquad\square$

The next theorem is a complete analogue of Theorem 1.2.4.

Theorem 7.2.5.

For a compact space K the following assertions are equivalent:

 (i) K is a Gul'ko compact;

 (ii) K continuously and injectively embeds into $(c_1(\Sigma' \times \Gamma), p)$ for some $\emptyset \neq \Sigma' \subset \mathbb{N}^{\mathbb{N}}$ and some $\Gamma \neq \emptyset$;

 (iii) $C(K)$ is a Vašák space;

(iv) $(B_{C(K)^*}, w^*)$ *is a Gul'ko compact;*

(v) *K admits a separating and weakly σ-point finite family $\mathcal{W}$ consisting of open F_σ sets in K; that is, for any distinct $k_1, k_2 \in K$ there is $W \in \mathcal{W}$ such that $W \cap \{k_1, k_2\}$ is a singleton, and there are subfamilies $\mathcal{W}_n \subset \mathcal{W}$, $n = 1, 2, \ldots$, such that for every $W \in \mathcal{W}$ and for every $k \in K$ there is $m \in \mathrm{IN}$ such that $W \in \mathcal{W}_m$ and $\mathrm{ord}(k, \mathcal{W}_m) := \#\{U \in \mathcal{W}_m : k \in U\}$ is finite; and*

(vi) *There exist sets $\Gamma \neq \emptyset$, $\Gamma_n \subset \Gamma$, $n = 1, 2, \ldots$, and a continuous injective mapping Φ from K into $[0,1]^\Gamma$ (in fact into $\Sigma(\Gamma)$) with the property that for every $\gamma \in \Gamma$ and for every $k \in K$ there is $m \in \mathrm{IN}$ such that $\gamma \in \Gamma_m$ and $\mathrm{supp}\,\Phi(k) \cap \Gamma_m := \{\gamma' \in \Gamma_m : \Phi(k)(\gamma') \neq 0\}$ is a finite set.*

Proof. The proof of (i)$\Rightarrow$(ii) follows by combining Theorems 7.1.8 and 7.2.4. That (i)$\Leftrightarrow$(iii) is contained in Theorem 7.1.8. That (iv)$\Rightarrow$(i) is from Theorem 7.1.10(ii). The proof of (i)$\Rightarrow$(iv) can be obtained by putting together Theorems 7.1.8 and 7.1.9.

(ii)$\Rightarrow$(v). Assume, for simplicity, that K is a subspace of $(c_1(\Sigma' \times \Gamma), p)$ for some $\Sigma' \subset \mathrm{IN}^{\mathrm{IN}}$ and some Γ. For $\sigma \in \Sigma'$, $\gamma \in \Gamma$, and $0 \neq r \in \mathrm{IR}$, we put

$$W^r_{\sigma,\gamma} = \{k \in K : rk(\sigma,\gamma) > 1\}.$$

Clearly, the sets $W^r_{\sigma,\gamma}$ are open F_σ in K. Let $\{B_n\}$ be a countable basis for the topology of the (separable metrizable) space Σ'. Then put

$$\mathcal{W}^r_n = \{W^r_{\sigma,\gamma} : \sigma \in B_n, \gamma \in \Gamma\}$$

and

$$\mathcal{W} = \bigcup \{\mathcal{W}^r_n : 0 \neq r \text{ rational}, n = 1, 2, \ldots\}.$$

It is easy to see that the family $\mathcal{W}$ separates the points of K. Fix $W \in \mathcal{W}$ and $k \in K$; thus $W = W^r_{\sigma,\gamma}$ for some r, σ, γ. Then there is $m \in \mathrm{IN}$ such that $B_m \ni \sigma$ and $rk(\sigma',\gamma') \leq 1$ whenever $\sigma \neq \sigma' \in B_m$ and $\gamma' \in \Gamma$. Indeed, if this were not the case, then there would exist $\sigma \neq \sigma_m \in \Sigma'$, $\sigma_m \to \sigma$, and $\gamma_m \in \Gamma$ such that $rk(\sigma_m, \gamma_m) > 1$. However, this would contradict the definition of the space $c_1(\Sigma' \times \Gamma)$ since $\{\sigma, \sigma_1, \sigma_2, \ldots\}$ is a compact set in Σ'. Then, clearly, $\mathrm{ord}(k, \mathcal{W}^r_m)$ is finite and $W \in \mathcal{W}^r_m$. It now remains to enumerate the couples (r,n) by the positive integers.

(v)$\Rightarrow$(vi). Let $\mathcal{W}$ be as in (v). Of course, such a $\mathcal{W}$ must be nonempty since it separates the points of K. For every $W \in \mathcal{W}$ we find a continuous function $f_W : K \to [0,1]$ such that $f_W(k) > 0$ if and only if $k \in W$ [En, Theorems 3.1.9 and 1.5.12]. We then define a mapping $\Phi : K \to [0,1]^{\mathcal{W}}$ as

$$\Phi(k)(W) = f_W(k), \qquad W \in \mathcal{W}, \qquad k \in K.$$

Clearly, Φ is continuous. It is also injective since $\mathcal{W}$ separates the points of K. Finally, take any $W \in \mathcal{W}$ and any $k \in K$ and let us find a corresponding $m \in \text{IN}$ as in (v). Then $W \in \mathcal{W}_m$ and

$$
\begin{aligned}
\text{supp } \Phi(k) \cap \mathcal{W}_m &= \{U \in \mathcal{W}_m : \ \Phi(k)(U) \neq 0\} \\
&= \{U \in \mathcal{W}_m : \ f_U(k) \neq 0\} = \{U \in \mathcal{W}_m : k \in U\},
\end{aligned}
$$

and the cardinality of the last set is, by (v), finite.

(vi)$\Rightarrow$(i). Let Γ, Γ_n and Φ be as in (vi). For $\gamma \in \Gamma$ and $k \in K$, put $\pi_\gamma(k) = \Phi(k)(\gamma)$; note that $\pi_\gamma \in C(K)$. Denote then

$$
Y = \{\pi_\gamma : \gamma \in \Gamma\}.
$$

This set separates the points of K since Φ is injective. Assertion (i) will be proved once we show that the space (Y, p) is $\mathcal{K}$-c.d.; see Theorem 7.1.8. We remark that (Y, p) is a subspace of the compact $([0,1]^K, p)$. For $n = 1, 2, \ldots$ we put

$$
A_n = \overline{\{\pi_\gamma : \ \gamma \in \Gamma_n\}}^p.
$$

(Here the p-closure is understood in $[0,1]^K$.) According to Definition 7.1.2 and Proposition 7.1.1(i), it is enough to show that for every $y \in Y$ and for every $f \in [0,1]^K \setminus Y$ there exists $n \in \text{IN}$ such that $y \in A_n$ and $f \notin A_n$. So fix such y and f. Then surely $f \not\equiv 0$, so there is some $k \in K$ satisfying $f(k) > 0$. We find $\gamma \in \Gamma$ so that $y = \pi_\gamma$. Using (vi), we find $m \in \text{IN}$ such that $\gamma \in \Gamma_m$ and the set supp $\Phi(k) \cap \Gamma_m$ is finite. Then surely $y \in A_m$. Finally, remarking that

$$
A_m = \{\pi_\gamma : \ \gamma \in \text{supp } \Phi(k) \cap \Gamma_m\} \cup \overline{\{\pi_\gamma : \ \gamma \in \Gamma_m \setminus \text{supp } \Phi(k)\}}^p,
$$

we conclude that $f \notin A_m$. $\qquad\qquad\square$

For a nonempty set Γ we define

$$
\Sigma(\Gamma) = \{f \in \text{IR}^\Gamma : \text{supp } f \text{ is at most countable}\}
$$

and endow it with the pointwise topology inherited from IR^Γ.

Definition 7.2.6. A Banach space V is called *weakly Lindelöf determined (WLD)* if (V^*, w^*) injects continuously into $(\Sigma(\Gamma), p)$ for some nonempty set Γ. A compact space is called *Corson compact* if it is homeomorphic to a subspace of $(\Sigma(\Gamma), p)$ for some set Γ.

We note that, owing to Theorems 1.2.5 and 1.2.4, WCG spaces are WLD and Eberlein compacta are Corson compacta. We have in fact a more general result:

Theorem 7.2.7.

Vašák spaces are WLD and Gul'ko compacta are Corson.

Proof. As regards to the first statement, it is enough to show, according to Theorem 7.2.4, that every element of $c_1(\Sigma' \times \Gamma)$, where $\Sigma' \subset \text{IN}^{\text{IN}}$, has countable support. Fix any f in this space and any $\epsilon > 0$. We observe that for every $\sigma \in \Sigma'$ there is a neighborhood W_σ of σ such that $f((W_\sigma \backslash \{\sigma\}) \times \Gamma) \subset (-\epsilon, \epsilon)$. Indeed, if this is not the case, then there would be $\sigma_n \in \Sigma'$, $\sigma_n \neq \sigma$ (Σ' is a metrizable space.), and $\gamma_n \in \Gamma$ such that $|f(\sigma_n, \gamma_n)| \geq \epsilon$ for all $n \in \text{IN}$. Put $L = \{\sigma, \sigma_1, \sigma_2, \ldots\}$. This is a compact set and hence $f_{|L \times \Gamma} \in c_0(L \times \Gamma)$, which is not true. Now, it is easy to check that Σ' is a Lindelöf space [En, Theorem 3.8.1]. Hence there exists a sequence $\{\tau_n\}$ in Σ' such that $\bigcup_{n=1}^{\infty} W_{\tau_n} = \Sigma'$. Now, if $|f(\sigma, \gamma)| > \epsilon$, then necessarily $\sigma = \tau_n$ for some $n \in \text{IN}$. We also know that $f_{|\{\sigma\} \times \Gamma} \in c_0(\{\sigma\} \times \Gamma)$, so there are only finitely many $\gamma' \in \Gamma$ with $|f(\sigma, \gamma')| > \epsilon$. We have just proved that the set $\{(\sigma, \gamma) \in \Sigma' \times \Gamma : |f(\sigma, \gamma)| > \epsilon\}$ is at most countable. Therefore $\text{supp} f$ is also at most countable.

Let K be a Gul'ko compact; so there are Γ, Γ_m, and Φ as in Theorem 7.2.5(vi). Fix any $k \in K$. For every $\gamma \in \Gamma$ we find $m \in \text{IN}$ such that $\gamma \in \Gamma_m$ and $\text{supp} \Phi(k) \cap \Gamma_m$ is finite. It then follows that $\text{supp} \Phi(k)$ is at most countable.

Another proof proceeds as follows. If K is Gul'ko, $C(K)$ is Vašák; thus, by Theorem 7.2.4, $(C(K)^*, w^*)$ injects into $(c_1(\Sigma' \times \Gamma), p)$. Therefore K is a Corson compact. $\qquad\square$

The next theorem generalizes and strengthens Theorem 4.1.2.

Theorem 7.2.8.

Let V be a Vašák space. Then $V \in \widetilde{\mathcal{F}}$, that is, (B_{V^*}, w^*) is fragmentable and hence V is a weak Asplund space.

Proof. By Theorem 7.1.9, (B_{V^*}, w^*) is a Gul'ko compact. Let $W = \bigcup_{n=1}^{\infty} W_n$ be the family found in Theorem 7.2.5(v) for (B_{V^*}, w^*). Define the sets

$$X_{mn} = \{v^* \in B_{V^*} : \text{ord}(v^*, W_n) \leq m\}, \qquad m, n \in \text{IN}.$$

It is easy to check that these sets are weak* closed.

Fix any $m, n \in \text{IN}$. We shall construct a relatively open partitioning $\mathcal{U}_{mn}$ of (B_{V^*}, w^*); see Definition 5.1.1. This will serve for fragmentability; see Theorem 5.1.9. For the first two steps, we set $U_0^{mn} = \emptyset$ and $U_1^{mn} = B_{V^*} \backslash X_{mn}$. Let $\xi > 1$ be a given ordinal and let us assume we have already constructed U_η^{mn} for every $\eta \in [0, \xi)$; put $W_\xi^{mn} = \bigcup_{\eta < \xi} U_\eta^{mn}$. If $B_{V^*} \backslash W_\xi^{mn}$ is an empty set, define $\mathcal{U}_{mn} = \{U_\eta^{mn} : \eta \in [0, \xi)\}$ and stop the process. Otherwise, we find $v_0^* \in B_{V^*} \backslash W_\xi^{mn}$ such that

$$\mathrm{ord}\left(v_0^*, \mathcal{W}_n\right) = \max\left\{\mathrm{ord}\left(v^*, \mathcal{W}_n\right): \ v^* \in B_{V^*} \backslash W_\xi^{mn}\right\}.$$

Such a v_0^* exists because $\xi > 1$ and hence $B_{V^*} \backslash W_\xi^{mn}$ is contained in X_{mn}. Then, putting

$$U_\xi^{mn} = \bigcap\left\{W \in \mathcal{W}_n: \ v_0^* \in W\right\} \backslash W_\xi^{mn}, \qquad (*)$$

this set contains v_0^* and is relatively open in $B_{V^*} \backslash W_\xi^{mn}$ since $\mathrm{ord}\left(v_0^*, \mathcal{W}_n\right)$ is finite. This finishes the construction of the relatively open partitioning $\mathcal{U}_{mn}$ of B_{V^*}.

It remains to show that the family $\mathcal{U} := \bigcup\left\{\mathcal{U}_{mn}: \ m, n \in \mathbb{N}\right\}$ separates the points of B_{V^*}. Let v_1^*, v_2^* be two distinct points of B_{V^*}. Since $\mathcal{W}$ separates the points of V^*, there is $W \in \mathcal{W}$ such that $\left\{v_1^*, v_2^*\right\} \cap W$ is a singleton, say $v_1^* \in W$ and $v_2^* \notin W$. Further, we know that there is $n \in \mathbb{N}$ such that $W \in \mathcal{W}_n$ and $m := \mathrm{ord}\left(v_1^*, \mathcal{W}_n\right)$ is finite. Thus $v_1^* \in X_{mn}$ and $v_1^* \in U_\xi^{mn}$ for some $\xi > 1$; see $(*)$. Hence $v_1^* \in W \cap U_\xi^{mn}$. Now the construction of U_ξ^{mn} guarantees that $W \supset U_\xi^{mn}$. Thus $v_2^* \notin U_\xi^{mn}$. This shows that the family $\mathcal{U}$ separates the points of B_{V^*}. Now we are ready to apply Theorem 5.1.9. $\square$

By putting together Theorems 7.1.8, 7.2.8, and 5.1.12, we get:

Theorem 7.2.9.

Every Gul'ko compact is fragmentable by a complete metric yielding a stronger topology, is sequentially compact, and contains a dense G_δ completely metrizable subset.

7.3. A NON-VAŠÁK SPACE WHOSE DUAL HAS A STRICTLY CONVEX DUAL NORM

Such a space was found by Johnson and Lindenstrauss [JL1, JL2; Di1, pp. 189, 190]. They constructed a non-WCG space V with V^* WCG. Then V is Asplund [Ph, Theorem 2.43]; in fact, V^* even has a dual locally uniformly rotund norm [DGZ2, Theorem VII.2.3.(ii)]. If V were Vašák, then, by Theorem 8.3.3, V would be WCG.

In what follows we present a different example in this vein. It will be of the type $C(K)$.

CONSTRUCTION OF THE COMPACT K

We shall first find a family $\{N_\xi: \ 1 \le \xi < \omega_1\}$ of infinite subsets of $\mathbb{N}$ which are mutually almost disjoint, that is, $N_\xi \cap N_\eta$ is a finite set whenever $1 \le \xi < \eta < \omega_1$. It is easy to choose infinite sets $N_i \subset \mathbb{N}$, $i \in \mathbb{N}$, such that

$N_i \cap N_j = \emptyset$ whenever $1 \le i < j < \omega$. Now let $\omega \le \xi < \omega_1$ be given and assume that we have already constructed infinite and mutually almost disjoint sets N_ζ, $\zeta < \xi$. Let $\{\zeta_1^\xi, \zeta_2^\xi, \ldots\}$ be an enumeration of the (countable) interval $[1, \xi)$. Let F_1 be any set in $N_{\zeta_1^\xi}$ with $\#F_1 = 1$. For $n = 2, 3, \ldots$ we find a set $F_n \subset N_{\zeta_n^\xi} \setminus (N_{\zeta_1^\xi} \cup \cdots \cup N_{\zeta_{n-1}^\xi})$, with $\#F_n = n$. Put then $N_\xi = F_1 \cup F_2 \cup \cdots$. We observe that $F_n \subset N_\xi \cap N_{\zeta_n^\xi} \subset F_1 \cup \cdots \cup F_n$, which is a finite set.

Having all N_ξ constructed, we put

$$\Phi(\eta, \xi) = \#(N_\eta \cap N_\xi) \quad \text{if} \quad 1 \le \eta < \xi < \omega_1.$$

Let $\psi : [0, \omega_1) \to [0, 1]$ be an injective mapping; it exists since $[0, 1]$ is an uncountable set. We define $\mathcal{A}_0$ as the family of all singletons from $[1, \omega_1)$ and all finite sets $\{\xi_1, \ldots, \xi_n\} \subset [1, \omega_1)$ such that $\xi_1 < \cdots < \xi_n$ and

$$\Phi(\xi_i, \xi_j) \ge j \quad \text{and} \quad |\psi(\xi_i) - \psi(\xi_j)| \le \frac{1}{i} \quad \text{whenever} \quad 1 \le i < j \le n.$$

Denote then

$$K = \mathrm{cl}\{\chi_A : A \in \mathcal{A}_0\}$$

(Here χ_A means the characteristic function of A.) and

$$\mathcal{A} = \{A \subset [1, \omega_1) : \chi_A \in K\}.$$

A function $f : X \to \mathbb{R}$ defined on a topological space is called *Baire class* 1 if there are continuous functions $f_n : X \to \mathbb{R}$, $n \in \mathbb{N}$, such that $f(x) = \lim_{n \to \infty} f_n(x)$ for every $x \in X$.

Definition 7.3.1. A compact space is called a *Rosenthal* compact if it is homeomoprhic to a compact subset of the space of Baire class 1 functions defined on a complete separable metric space endowed with the pointwise topology.

Theorem 7.3.2.

The space K is Corson and Rosenthal compact but not Gul'ko compact; hence $C(K)$ is not a Vašák space.

Proof. Let $A \in \mathcal{A}$. If A is finite, then it necessarily belongs to $\mathcal{A}_0$. Let further A be infinite. We find $\xi_1 < \xi_2 < \cdots < \omega_1$ such that $\xi_i \in A$ and $\#(A \cap [0, \xi_i]) = i$, $i \in \mathbb{N}$. Put $\xi = \lim_{i \to \infty} \xi_i$; then $\xi < \omega_1$. Assume that there is $\eta \in A \cap [\xi, \omega_1)$. We recall that $\chi_A = \lim_\tau \chi_{F_\tau}$, where $\{F_\tau\}$ is a net in $\mathcal{A}_0$. Hence for every $i \in \mathbb{N}$ the set $\{\xi_1, \ldots, \xi_i, \eta\}$ belongs to $\mathcal{A}_0$. Thus $\Phi(\xi_i, \eta) \ge i + 1$ for all $i \in \mathbb{N}$, which is impossible. Therefore A is of the form $\{\xi_1, \xi_2, \ldots\}$ and K is a Corson compact.

Define a mapping $\Psi : K \rightarrow [0,1]^{[0,1]}$ by $\Psi(\chi_A) = \chi_{\psi(A)}$, $\chi_A \in K$. It is elementary to check that Ψ is injective and continuous. Take any $\chi_A \in K$. If A is finite, then clearly, $\chi_{\psi(A)}$ is a limit of a sequence of continuous functions on $[0,1]$. If A is infinite, we can easily check that $\psi(A)$ is a convergent sequence. Hence, again, $\chi_{\psi(A)}$ is a Baire class 1 function on $[0,1]$. Therefore $\Psi(K)$, and hence K, is a Rosenthal compact.

Take a point $*$ outside of $[1,\omega_1)$, put $\Gamma_* = [1,\omega_1) \cup \{*\}$ and endow it with the following topology: Every $\xi \in [1,\omega_1)$ is isolated and a subbase of neighborhoods of $*$ is $\{\Gamma_* \backslash A : A \in \mathcal{A}\}$. Consider the mapping $\delta : \Gamma_* \rightarrow (C(K), p)$ defined by $\delta(\xi)(k) = k(\xi)$ if $\xi \in [1,\omega_1)$ and $\delta(*)(k) = 0$ for all $k \in K$. It is elementary to verify that δ is a homeomorphism of Γ_* onto $(\delta(\Gamma_*), p)$. We further observe that $\delta(\Gamma_*)$ is closed in $(C(K), p)$. So let $\{\delta(\xi_\tau)\}$ be a net converging pointwise to an $f \in C(K)$. Assume first that $f(\chi_{\{\xi\}}) = 1$ for some $\xi \in [1,\omega_1)$. Then $\xi_\tau = \xi$ for all large τ and hence $f = \delta(\xi) \in \delta(\Gamma_*)$. Second, assume that $f(\chi_{\{\xi\}}) = 0$ for all $\xi \in [0,\omega_1)$. (There is no third possibility.) Then

$$f(\chi_F) = \lim_\tau \delta(\xi_\tau)(\chi_F) = \lim_\tau \chi_F(\xi_\tau)$$

$$= \sum_{\eta \in F} \lim_\tau \chi_{\{\eta\}}(\xi_\tau) = \sum_{\eta \in F} \lim_\tau \delta(\xi_\tau)(\chi_{\{\eta\}})$$

$$= \sum_{\eta \in F} f(\chi_{\{\eta\}}) = 0$$

for every $F \in \mathcal{A}_0$. Now, if $A \in \mathcal{A}$ is infinite, we know from the first paragraph that there are sets $F_i \in \mathcal{A}_0$ such that $F_1 \subset F_2 \subset \cdots \subset A$ and $\bigcup_{i=1}^\infty F_i = A$. Thus, $f(\chi_A) = \lim_{i \rightarrow \infty} f(\chi_{F_i}) = 0$. This means that f is equal to 0 identically; so $f = \delta(*) \in \delta(\Gamma_*)$.

Assume for the purpose of obtaining a contradiction that K is a Gul'ko compact, that is, $(C(K), p)$ is $\mathcal{K}$-c.d. Then, by Theorem 7.1.3, $\delta(\Gamma_*)$ and hence Γ_* are also $\mathcal{K}$-c.d. This means, by Proposition 7.1.1, that there are a compact space L, closed sets $B_s \subset L$, $s \in S$ ($=$ the set of all finite sequences of positive integers), and $\Sigma' \subset \mathbb{N}^{\mathbb{N}}$ such that Γ_* is a subspace of L and

$$\Gamma_* = \bigcup_{\sigma \in \Sigma'} \bigcap_{n=1}^\infty B_{\sigma|n}.$$

For $s \in S$, put $C_s = B_{s|1} \cap \cdots \cap B_{s|n} \cap \Gamma_*$, where $n = |s|$.

We shall say that a subset of $[1,\omega_1)$ is *stationary* if it intersects every closed unbounded (i.e., uncountable) subset of $[1,\omega_1)$. It is easy to check that the intersection of countably many closed unbounded sets from $[1,\omega_1)$ is nonempty and even unbounded. In particular, closed unbounded sets are stationary. We put $S_1 = \{s \in S : C_s \text{ is not stationary}\}$ and $C = \bigcup \{C_s : s \in S_1\}$. For every $s \in S_1$ there is a closed unbounded set $A_s \subset [1,\omega_1)$ such that $C_s \cap A_s = \emptyset$. Then $[1,\omega_1) \backslash C \supset \bigcap_{s \in S_1} A_s$ and so $[1,\omega_1) \backslash C$ is an unbounded set. For $s \in S \backslash S_1$ define $f^s : C_s \rightarrow [1,\omega_1)$ as $f^s(\xi) = \max\{\zeta_1^\xi, \ldots, \zeta_n^\xi\}$, $\xi \in C_s$, where ζ_i^ξ are from the construction of K and $n = |s|$. Since $f^s(\xi) < \xi$, Fodor's theorem [Fo] yields a stationary set $E_s \subset C_s$ and $\eta^s \in [1, \omega_1)$ such that $f^s(\xi) = \eta^s$ for each $\xi \in E_s$.

We choose $\xi \in [1, \omega_1) \backslash C$ with $\xi > \sup\{\eta^s : s \in S \backslash S_1\}$. We find $\sigma \in \Sigma'$ such that $\xi \in \bigcap_{n=1}^{\infty} B_{\sigma|n} (= \bigcap_{n=1}^{\infty} C_{\sigma|n})$; thus $\sigma|n \in S \backslash S_1$ for every $n \in \mathbb{N}$. We take $\xi < \xi_1 < \xi_2 < \cdots < \omega_1$ such that $\xi_n \in E_{\sigma|n}$, $n \in \mathbb{N}$. We shall show that $\{\xi_{n_1}, \xi_{n_2}, \ldots\} \in \mathcal{A}$ for some $n_1 < n_2 < \cdots$. Take $1 \le i < j < \omega$. Then $[1, \xi_j) = \{\zeta_1^{\xi_j}, \zeta_2^{\xi_j}, \ldots\}$ (We use the notation from the construction of K.) and

$$\max\{\zeta_1^{\xi_j}, \ldots, \zeta_j^{\xi_j}\} = f^{\sigma|j}(\xi_j) = \eta^{\sigma|j} < \xi < \xi_i < \xi_j.$$

Hence $\xi_i = \zeta_l^{\xi_j}$ with $l > j$. Thus, from the construction of N_{ξ_j}, we can find $F_l \subset N_{\xi_j} \cap N_{\xi_i}$, with $\#F_l = l$. Therefore

$$\Phi(\xi_i, \xi_j) = \#(N_{\xi_i} \cap N_{\xi_j}) \ge l > j.$$

Now let x be a cluster point of the sequence $\{\psi(\xi_n)\}$; so there are $n_1 < n_2 < \cdots$ such that $|\psi(\xi_{n_i}) - x| < 1/(2i)$, $i \in \mathbb{N}$. Then

$$|\psi(\xi_{n_i}) - \psi(\xi_{n_j})| < \frac{1}{i} \quad \text{whenever} \quad 1 \le i < j < \omega.$$

It then follows that $\{\xi_{n_1}, \ldots, \xi_{n_j}\} \in \mathcal{A}_0$ for all $j \in \mathbb{N}$, that is, $\{\xi_{n_1}, \xi_{n_2}, \ldots\} \in \mathcal{A}$.

Finally, we recall that $\xi_n \in E_{\sigma|n} \subset C_{\sigma|n} = B_{\sigma|1} \cap \cdots \cap B_{\sigma|n} \cap \Gamma_* \subset L$. Let $\gamma \in L$ be a cluster point of the sequence $\{\xi_{n_i}\}$. Then $\gamma \in \bigcap_{n=1}^{\infty} B_{\sigma|n} \subset \Gamma_*$ and, consequently, γ must be equal to $*$. However, this is impossible since $\Gamma_* \backslash \{\xi_{n_1}, \xi_{n_2}, \ldots\}$ is a neighborhood of $*$. This contradiction means that K is not Gul'ko compact. $\qquad\square$

Let $M(K)$ denote the set of all regular Borel measures on K. Let $\|\mu\|$ mean the total variation of $\mu \in M(K)$. In what follows, we shall identify, using F. Riesz's representation theorem, the Banach space $(M(K), \|\cdot\|)$ with the Banach space $C(K)^*$ endowed with the norm dual to the supremum norm on $C(K)$. If $L \subset K$ is a Borel set and $\mu \in M(K)$, we put $\mu_{|L}(A) = \mu(A \cap L)$ for all Borel sets $A \subset K$; so $\mu_{|L} \in M(K)$.

Lemma 7.3.3.

Let K be a compact space and let $L \subset K$ be a closed subset such that $C(L)^*$ admits an equivalent strictly convex dual norm. Then $C(K)^*$ admits an equivalent dual norm $\|\|\cdot\|\|$ such that $\mu_{1|L} = \mu_{2|L}$ whenever $\mu_1, \mu_2 \in C(K)^*$ and $\frac{1}{2}\|\|\mu_1 + \mu_2\|\| = \|\|\mu_1\|\| = \|\|\mu_2\|\|$.

Proof. Denote $M_0 = \{\mu_{|L} : \mu \in C(K)^*\}$ and define $Q : M_0 \to C(L)^*$ as

$$\langle Q\mu, f \rangle = \int_L f \, d\mu, \qquad \mu \in M_0, \qquad f \in C(L).$$

We observe that Q is an isometry. Indeed, for $\mu \in M_0$ we have

$$\|Q\mu\| = \sup\langle Q\mu, B_{C(L)}\rangle = \sup\left\{ \int_L f \, d\mu : f \in B_{C(L)}\right\}$$
$$= \sup\left\{ \int_K g \, d\mu : g \in B_{C(K)}\right\} = \|\mu\|;$$

here we used the Tietze–Urysohn extension theorem and the definition of integral. The mapping Q is also weak*-to-weak* continuous for if μ_τ, $\mu \in M_0$ and $\mu_\tau \to \mu$ weak*, then for every $f \in C(L)$

$$\langle Q\mu_\tau, f\rangle = \int_L f \, d\mu_\tau = \int_K g \, d\mu_\tau \to \int_K g \, d\mu = \int_L f \, d\mu,$$

where $g \in C(K)$ was such that $g_{|L} = f$; see again the Tietze–Urysohn theorem.

We shall now show that M_0 is weak* closed. Let $\{\mu_\tau\} \subset M_0$ be a net weak* converging to some $\mu \in C(K)^*$. According to another Krein–Šmulyan theorem [DS, Theorem V.5.7], we may and do assume that $\{\mu_\tau\}$ is bounded. Then $\{Q\mu_\tau\}$ is a bounded net in $C(L)^*$ and hence there is its subnet weak* converging to a $\lambda \in C(L)^*$. We put $\nu(A) = \lambda(A \cap L)$ for every Borel set $A \subset K$; then $\nu \in M(K)$ and $\nu_{|L} = \nu$, so $\nu \in M_0$. Now, for any $f \in C(K)$ we have

$$\langle \mu_\tau, f\rangle = \int_L (f_{|L}) \, d\mu_\tau = \langle Q\mu_\tau, f_{|L}\rangle.$$

Consequently,

$$\langle \mu, f\rangle = \lim_\tau \langle \mu_\tau, f\rangle = \lim_\tau \langle Q\mu_\tau, f_{|L}\rangle = \langle \lambda, f_{|L}\rangle$$
$$= \int_L (f_{|L}) \, d\lambda = \int_K f \, d\nu = \langle \nu, f\rangle$$

for every $f \in C(K)$ and therefore $\mu = \nu \in M_0$.

Let $|\cdot|$ be an equivalent dual strictly convex norm on $C(L)^*$. We may and do assume that $|\cdot| \leq \|\cdot\|$, where $\|\cdot\|$ is the norm dual to the supremum norm on $C(L)$. Define

$$\|\|\mu\|\| = |Q(\mu_{|L})| + \|\mu_{|(K\setminus L)}\|, \qquad \mu \in C(K)^*.$$

Clearly, $\|\|\cdot\|\|$ is an equivalent norm on $C(K)^*$. We even have

$$\|\|\mu\|\| \leq \|Q(\mu_{|L})\| + \|\mu_{|(K\setminus L)}\| = \|\mu_{|L}\| + \|\mu_{|(K\setminus L)}\| = \|\mu\|.$$

We shall prove that $\|\|\cdot\|\|$ is a dual norm, that is, that its unit ball is weak* closed. So take a net $\{\mu_\tau\} \subset C(K)^*$, with $\|\|\mu_\tau\|\| \leq 1$, weak* converging to some $\mu \in C(K)^*$. It is enough to show that $\|\|\mu\|\| \leq 1$. Let λ be a weak* cluster point of the net $\{\mu_{\tau|L}\}$. We note that $\lambda \in M_0$ since M_0 was shown to be weak* closed; so $\||\lambda\|| = |Q(\lambda)|$. For brevity, we shall assume that $\mu_{\tau|L} \to \lambda$ weak*. Then $Q(\mu_{\tau|L}) \to Q(\lambda)$ weak*, $\mu_{\tau|(K\setminus L)} \to \mu - \lambda$ weak*, and so

$$|Q\lambda| \le \liminf_\tau |Q(\mu_{\tau|L})| \quad \text{and} \quad \|\mu - \lambda\| \le \liminf_\tau \|\mu_{\tau|(K\setminus L)})\|$$

since $|\cdot|$ and $\|\cdot\|$ are weak* lower semicontinuous. Now we can estimate

$$\begin{aligned}
\||\mu\|| &\le \||\lambda\|| + \||\mu - \lambda\|| \le \||\lambda\|| + \|\mu - \lambda\| \\
&= |Q\lambda| + \|\mu - \lambda\| \le \liminf_\tau |Q(\mu_{\tau|L})| + \liminf_\tau \|\mu_{\tau|(K\setminus L)}\| \\
&\le \liminf_\tau \left(|Q(\mu_{\tau|L})| + \|\mu_{\tau|(K\setminus L)}\| \right) = \liminf_\tau \||\mu_\tau\|| \le 1.
\end{aligned}$$

Therefore $\||\cdot\||$ is a dual norm on $C(K)^*$.

Finally, let $\mu_1, \mu_2 \in C(K)^*$ be such that $\frac{1}{2}\||\mu_1 + \mu_2\|| = \||\mu_1\|| = \||\mu_2\||$. From convexity we obtain that $\frac{1}{2}|Q((\mu_1 + \mu_2)_{|L})| = |Q(\mu_{1|L})| + |Q(\mu_{2|L})|$. Since $|\cdot|$ is strictly convex, we have $Q(\mu_{1|L}) = Q(\mu_{2|L})$ and therefore $\mu_{1|L} = \mu_{2|L}$. $\qquad\square$

Let K mean the compact constructed at the beginning of this section.

Theorem 7.3.4.

The dual space $C(K)^*$ admits an equivalent strictly convex dual norm and hence $C(K)$ is a weak Asplund space. Moreover, $C(K)$ is WLD.

Proof. For any finite set $F \subset [1, \omega_1)$ we put

$$W_F = \{\chi_A \in K : A \supset F\}.$$

Clearly, such sets are closed and open in K, so $\chi_{W_F} \in C(K)$ for every finite set $F \subset [1, \omega_1)$. For $m = 2, 3, \ldots$ put

$$D_m = \left\{ \chi_{W_F} : F \subset [1, \omega_1), \ 2 \le \#F < \aleph_0, \ \max\{\Phi(\zeta, \xi) : \zeta, \xi \in F, \ \zeta < \xi\} = m \right\}$$

and $D = \{1\} \cup \bigcup_{m=2}^\infty D_m$, where 1 means the function on K which is identically equal to 1. We define a mapping $T : C(K)^* \to \ell_\infty(D)$ as follows:

$$T\mu(f) = \frac{1}{m}\langle \mu, f \rangle, \qquad \mu \in C(K)^* \quad \text{if} \quad f \in D_m \quad \text{and} \quad m \in \{2, 3, \ldots\}$$

and $T\mu(1) = \mu(K)$. We can immediately check that T is bounded, linear, and weak*-to-pointwise continuous.

We claim that $T(X) \subset c_0(D)$, where $X = \{\pm \delta_k : k \in K\}$ and δ_k means the point mass measure at k. So fix $k = \chi_A \in K$, $m \in \{2, 3, \ldots\}$ and consider any $f = \chi_{W_F} \in D_m$ such that $T\delta_k(f) \ne 0$. Then

$$0 \ne \langle \delta_k, f \rangle = f(k) = \chi_{W_F}(\chi_A);$$

hence $\chi_A \in W_F$, that is, $F \subset A$. If A is finite, then we clearly have just finitely many possibilities for F, that is, for f. Second, let A be infinite. From the proof of Theorem 7.3.2 we know that A is of the form $\{\xi_1, \xi_2, \ldots\}$, where $\xi_1 < \xi_2 < \cdots < \omega_1$. Therefore F is of the form $\{\xi_{n_1}, \xi_{n_2}, \ldots, \xi_{n_q}\}$, with $n_1 < n_2 < \cdots < n_q$. But $f \in D_m$ and so $\Phi(\xi_{n_1}, \xi_{n_q}) \le m$. Also, since $\{\xi_1, \xi_2, \ldots, \xi_{n_q}\} \in \mathcal{A}_0$, we have $\Phi(\xi_{n_1}, \xi_{n_q}) \ge n_q$, and so $n_q \le m$. Therefore only finitely many $f \in D_m$ satisfy $T\delta_k(f) \ne 0$. This finishes the proof of the claim.

We further claim that $T(B_{C(K)^*})$, and hence $T(C(K)^*)$, lie in $c_0(D)$. Indeed, a simple application of the separation theorem and [DS, Theorem V.3.9] yields that $B_{C(K)^*} = \overline{\text{co}X}^*$. Then the weak*-to-pointwise continuity of T guarantees that

$$T(B_{C(K)^*}) = T(\overline{\text{co}X}^*) \subset \overline{T(\text{co}X)}^p = \overline{\text{co}(T(X))}^p,$$

where p means the pointwise topology. Also, as X is weak* compact, $T(X)$ is pointwise compact in $c_0(D)$ and so $T(X)$ is in fact weakly compact, as T is bounded. Now, the Krein–Šmulyan theorem guarantees that $\overline{\text{co}(T(X))}$ is weakly and hence pointwise compact. Therefore $\overline{\text{co}(T(X))}^p = \overline{\text{co}(T(X))} \subset c_0(D)$.

Put $L = \{\chi_{\{\xi\}} : \xi \in [1, \omega_1)\} \cup \{0\}$. Trivially, 0 is the only accumulation point of L. It then follows that $C(L)$ is isomorphic with $c_0(L) \times \mathbb{R}$, the last space being Asplund and WCG. Hence $C(L)^*$ has an equivalent locally uniformly rotund (hence strictly convex) dual norm [DGZ2, Theorem II.2.1]. Let $\|\|\cdot\|\|$ be the norm on $C(K)^*$ constructed for our L in Lemma 7.3.3.

Define

$$n(\mu) = \|\|\mu\|\| + \|T\mu\|, \qquad \mu \in C(K)^*,$$

where $\|\cdot\|$ is Day's norm on $c_0(D)$ [DGZ2, p. 69]. Here, n is clearly an equivalent norm. It is also a dual norm since $\|\|\cdot\|\|$ is and T is weak*-to-weak continuous. It remains to show that n is strictly convex. So consider $\mu_1, \mu_2 \in C(K)^*$ satisfying

$$\tfrac{1}{2}n(\mu_1 + \mu_2) = n(\mu_1) = n(\mu_2).$$

From convexity, we have

$$\tfrac{1}{2}\|\|\mu_1 + \mu_2\|\| = \|\|\mu_1\|\| = \|\|\mu_2\|\| \quad \text{and} \quad \tfrac{1}{2}\|T(\mu_1 + \mu_2)\| = \|T\mu_1\| = \|T\mu_2\|.$$

Then Lemma 7.3.3 and the strict convexity of $\|\cdot\|$ [DGZ2, Theorem II.7.3] yield

$$\mu_{1|L} = \mu_{2|L} \quad \text{and} \quad T\mu_1 = T\mu_2.$$

Denote $\nu = \mu_1 - \mu_2$. Then $\nu_{|L} = 0$, $\nu(K) = 0$, and $\nu(W_F) = \langle \nu, \chi_{W_F} \rangle = 0$ for every $F \subset [1, \omega_1)$ with $2 \leq \#F < \aleph_0$. Fix any $\xi \in [1, \omega_1)$; then

$$W_{\{\xi\}} = \{\chi_{\{\xi\}}\} \cup \bigcup \{W_F : \xi \in F \subset [1, \omega_1), \ 2 \leq \#F < \aleph_0\}$$

and the regularity of ν guarantees that $\nu(W_{\{\xi\}}) = \nu(\{\chi_{\{\xi\}}\})$. However, $\nu_{|L} = 0$ and $\chi_{\{\xi\}} \in L$. So $\nu(\{\chi_{\{\xi\}}\}) = 0$.

We thus have $\nu(W_F) = 0$ for all finite sets $F \subset [1, \omega_1)$. Denote

$$H = \{\chi_{W_F} : \ F \subset [1, \omega_1), \ 0 \leq \#F < \aleph_0\}.$$

This set clearly separates the points of K and contains the function which is identically equal to 1. We also observe that $\chi_{W_F} \cdot \chi_{W_G} = \chi_{W_F \cap W_G} = \chi_{W_{F \cup G}}$. It then follows, by the Stone–Weierstrass theorem, that $\overline{\mathrm{sp}}\, H = C(K)$. Now we know that $\langle \nu, f \rangle = 0$ for all $f \in H$, and hence also for all $f \in \mathrm{sp}\, H$. Therefore $\nu = 0$, that is, $\mu_1 = \mu_2$. We have thus proved that $C(K)^*$ admits an equivalent strictly convex dual norm and Theorems 4.2.3 and 3.2.2 guarantee that $C(K)$ is a weak Asplund space.

Since we know that K is a Rosenthal compact, it has a property (M), that is, every positive regular Borel measure on K has a separable support [Go1]. It now remains to apply the following lemma. $\square$

Lemma 7.3.5.

Let K be a Corson compact having the property that for every positive $\mu \in M(K)$ there exists a separable Borel set $B \subset K$ such that $\mu(B) = \mu(K)$. Then $C(K)$ is a WLD space.

Proof. Since K is a Corson compact, we may think that K is a subspace of $(\Sigma(\Gamma), p)$ for some set Γ. Let us consider the mapping $\psi : \mathrm{IR}^\Gamma \to [0, +\infty)^{\Gamma \times \{0,1\}}$ defined by

$$\psi(x)(\gamma, 0) = \max(x_\gamma, 0) \quad \text{and} \quad \psi(x)(\gamma, 1) = \max(-x_\gamma, 0),$$

$$\gamma \in \Gamma, \quad x = \{x_{\gamma'}\} \in \mathrm{IR}^\Gamma.$$

Then ψ is continuous and injective and maps $\Sigma(\Gamma)$ into $\Sigma(\Gamma \times \{0, 1\})$. It follows that we may and do assume that $x_\gamma \geq 0$ for every $x \in K$ and every $\gamma \in \Gamma$. Finally, by multiplying some coordinates, if necessary, we get that $K \subset [0, 1]^\Gamma$.

For $\gamma \in \Gamma$ we define $\pi_\gamma : K \to \mathrm{IR}$ by $\pi_\gamma(k) = k_\gamma$, $k = \{k_{\gamma'}\} \in K$. Clearly, $\pi_\gamma \in C(K)$. We can immediately observe that the family $\{\pi_\gamma : \ \gamma \in \Gamma\}$ is point countable, that is, for each $k \in K$ the set $\{\gamma \in \Gamma : \ \pi_\gamma(k) \neq 0\}$ is at most countable, and that it separates the points of K, that is, whenever $k, k' \in K$ and $k \neq k'$, then $\pi_\gamma(k) \neq \pi_\gamma(k')$ for some $\gamma \in \Gamma$. For $n = 1, 2, \ldots$ we put $\Delta_n = \{\pi_{\gamma_1} \cdot \pi_{\gamma_2} \cdots \pi_{\gamma_n} : \ \gamma_1, \gamma_2, \ldots, \gamma_n \in \Gamma\}$ and then $\Delta = \bigcup_{n=1}^\infty \Delta_n \cup \{\text{constant}$

function 1 on K}. Now we define $\Phi : (C(K)^*, w^*) \to [0, +\infty)^\Delta$ by

$$\Phi(\mu)(f) = \langle \mu, f \rangle, \qquad f \in \Delta, \qquad \mu \in C(K)^*.$$

Then Φ is clearly linear and weak*-to-pointwise continuous. The mapping Φ is also injective because, by the Stone–Weierstrass theorem, the set Δ is linearly dense in $C(K)$. We shall show that $\Phi(C(K)^*) \subset \Sigma(\Delta)$. So fix a $\mu \in C(K)^*$. Then μ can be written as $\mu = \mu_1 - \mu_2$, where μ_1, μ_2 are positive measures on K; see F. Riesz's theorem and [DS, Theorem III.4.10]. By the assumption, we know that there are separable sets $B_i \subset K$ such that $\mu_i(B_i) = \mu_i(K)$, $i = 1, 2$. Then, since K is in $\Sigma(\Gamma)$, there is a countable set $\Gamma_0 \subset \Gamma$ such that $x_\gamma = 0$ whenever $x = \{x_{\gamma'}\} \in B_i$ and $\gamma \in \Gamma \backslash \Gamma_0$. But then for every $\gamma \in \Gamma \backslash \Gamma_0$ we have

$$\langle \mu_i, \pi_\gamma \rangle = \int_K \pi_\gamma(x) \, \mathrm{d}\mu_i(x) = \int_{B_i} \pi_\gamma(x) \, \mathrm{d}\mu_i(x) = \int_{B_i} x_\gamma \, \mathrm{d}\mu_i(x) = 0$$

and hence the set $\{\gamma \in \Gamma : \langle \mu_i, \pi_\gamma \rangle \neq 0\}$ lies in Γ_0. Assume now that we have $n \in \mathrm{IN}$ and $\gamma_1, \gamma_2, \ldots, \gamma_n \in \Gamma$ such that $\langle \mu_i, \pi_{\gamma_1} \cdots \pi_{\gamma_n} \rangle \neq 0$. Since $\pi_\gamma \geq 0$ for all $\gamma \in \Gamma$, we have $\langle \mu_i, \pi_{\gamma_j} \rangle > 0$ for all $j = 1, 2, \ldots, n$. Therefore all such γ_j must belong to Γ_0. Hence $\{f \in \Delta : \langle \mu_i, f \rangle \neq 0\}$ is countable, $\{f \in \Delta : \langle \mu, f \rangle \neq 0\}$ is countable and finally $\Phi(C(K)^*) \subset \Sigma(\Gamma)$. This means that $C(K)$ is WLD; see Definition 7.2.6. $\qquad\square$

7.4. NOTES AND REMARKS

The concept of Vašák space is due to Vašák [V]; see also his predecessors [Ch1, Ch2, Fr]. Almost all facts presented in the first section and many other related results can be found in the fundamental paper of Talagrand [Ta2].

That (iv) in Theorem 7.1.10 cannot be extended to countable unions was shown by Sokolov [So]. He considered a compactification $b\mathrm{IN}$ of IN which is a countable union of Eberlein compacta and yet not Corson compact. For a positive result, see the same paper.

A comparison of Theorem 7.1.11 with Theorem 3.2.3 raises the following natural questions. *If $T : Y \to V$ is bounded linear with T^{**} injective and V Vašák, is Y Vašák? Are uncountable c_0 sums and ℓ_p sums, $1 < p < +\infty$, of Vašák spaces Vašák?* We can see, without much effort, that the second question has positive answer for weakly $\mathcal{K}$-analytic spaces.

Proposition 7.2.1 is due to Orihuela and Valdivia [OV]. That Vašák spaces admit a P.R.I. was first proved by Vašák [V]. He did so by elaborating a technique of Amir and Lindenstrauss [AL]. A different proof was presented by Gul'ko [Gu]. He proved a bit more: *If K is a Gul'ko compact, then $C(K)$ admits a P.R.I. $\{P_\alpha\}$ such that P_α are continuous with respect to the topology of pointwise convergence; thus $(C(K), p)$ embeds continuously and injectively into $(c_0(\Gamma), p)$.* Our approach is based on Gul'ko's ideas, further developed and simplified in [OV]; see Proposition 7.2.1 and Section 6.1. Another proof is due to Stegall [St8].

The definition of c_1-space and Theorem 7.2.4 is due to Mercourakis [Me]. Our proof of the necessity is from [FT]; see also [DGZ2, p. 252]. For still another proof, see [Po]. In Theorem 7.2.5 (i)⇔(ii) is from [Me]. The equivalences (i)⇔(v)⇔(vi) are originally due to Sokolov [So]. It should be noted that (i)⇔(v) in Theorem 7.2.5 resembles Rosenthal's well-known characterization of Eberlein compacta; see Theorem 1.2.4, [Ro], and [Di1, Chapter 5, §3]. That Gul'ko compacta are Corson was first proved by Gul'ko [Gu].

Theorems 7.2.8 and 7.2.9 are due to Ribarska [Ri1]; see also [K4, Gr]. Kenderov was the first to prove that Vašák spaces are weak Asplund [K4]. That Gul'ko compacta contain a completely metrizable dense G_δ set was first proved by Leiderman [Le]; see also Gruenhage [Gr]. There is another proof of Theorem 7.2.8: According to [Me], *if V is Vašák, then V^* admits an equivalent strictly convex dual norm.* Hence, by Theorem 5.3.1, (V^*, w^*) is fragmentable.

When speaking about renormings, we mention that *a (dual) Vašák space admits a (dual) LUR norm* [F4; DGZ2, Theorem VII.2.3] and that *if V^* is Vašák, then V admits an equivalent LUR norm such that its dual norm is also LUR* [DGZ2, Theorem VII.2.7; FT].

In Section 7.3, we followed the paper of Argyros and Mercourakis [AM]; see also [AMN]. First examples of Corson which are not Gul'ko compacta are due to Alster and Pol [AP] and also to Leiderman and Sokolov [LS]. A Corson compact containing no dense metrizable subset (hence not Gul'ko) was constructed by Todorčević [To] (requiring no special axioms of set theory).

We recall that a topological space Y is called $\mathcal{K}$-*analytic* if it is a subspace of a compact space and there exists a usco mapping from $\mathbb{N}^{\mathbb{N}}$ onto Y. A Banach space V is called *weakly $\mathcal{K}$-analytic* if (V, w) is $\mathcal{K}$-analytic. Finally, a compact space K is called a *Talagrand* compact if $(C(K), p)$ is $\mathcal{K}$-analytic. It should be noted that, if we replace Σ', $\mathcal{K}$-c.d., Vašák and Gul'ko by $\mathbb{N}^{\mathbb{N}}$, $\mathcal{K}$-analytic, weakly $\mathcal{K}$-analytic, and Talagrand, respectively, all the statements (together with their proofs) in Sections 7.1 and 7.2 remain valid with two exceptions: In Proposition 7.1.1, (i) should be dropped, and in Theorem 7.2.5, (v) and (vi) should be respectively replaced by the following:

(v) *The compact K admits a separating family $\mathcal{W}$ consisting of open F_σ sets in K and subfamilies $\mathcal{W}_s \subset \mathcal{W}$, $s \in S$ (S was defined at the beginning of Section 7.1.), such that $\mathcal{W}_\emptyset = \mathcal{W}$,*

$$\mathcal{W}_s = \bigcup_{n=1}^{\infty} \mathcal{W}_{s \cdot n} \quad \text{for all} \quad s \in S,$$

and whenever $k \in K$ and $\sigma \in \mathbb{N}^{\mathbb{N}}$, then $\mathrm{ord}\,(k, \mathcal{W}_{\sigma|n})$ is finite for all large $n \in \mathbb{N}$.

(vi) There are sets Γ, $\Gamma_s \subset \Gamma$, $s \in S$, *and a continuous injection* $\Phi : K \to [0,1]^{\Gamma}$ *such that* $\Gamma_{\emptyset} = \Gamma$,

$$\Gamma_s = \bigcup_{n=1}^{\infty} \Gamma_{s \cdot n} \quad \text{for all} \quad s \in S,$$

and whenever $k \in K$ *and* $\sigma \in \mathbb{N}^{\mathbb{N}}$, *then* $\text{supp}\, \Phi(k) \cap \Gamma_{\sigma|n}$ *is finite for all large* $n \in \mathbb{N}$.

See [LS].

There is a natural question if there exists a Vašák space which is not weakly $\mathcal{K}$-analytic. This was answered positively by Talagrand [Ta4]. His construction has recently been made more transparent by Čížek [Či].

Quite recently Kenderov and Moors characterized fragmentability by the existence of a winning strategy for one of the players of a Banach–Mazur-like game. In this way they were able to prove Theorem 7.2.8 without any use of P.R.I. [KM].

For more information about the topic of this chapter we refer the reader to [Ta2; V; Gu; Me; So; LS; Ne, Section 6; MN; AMN; AM; Arg; Ar, Chapter IV; DGZ2, Chapter VI].

Chapter Eight

A Characterization of WCG Spaces and of Eberlein Compacta

In the first section, we prove a selection theorem due to Jayne and Rogers [JR], and a well-known inequality due to Simons [Si]. In the second section, these tools are used in the construction of a P.R.I. in the dual of a general Asplund space. In the third section, we find equivalent characterizations of WCG Banach spaces and their subspaces. In particular, we show that a Vašák or even WLD space which is simultaneously (a subspace of) an Asplund generated space must already be (a subspace of) a WCG space. Topological counterparts of such equivalences are also presented here: If the continuous image of a Radon–Nikodým compact is Gul'ko (or even Corson) compact, then it is Eberlein compact. The last section contains a counterexample related to Theorem 8.3.3.

8.1. PREPARATORY FACTS

Let X and Y be topological spaces. A mapping $f: X \to Y$ is said to be *Baire class* 1 if there are continuous mappings $f_n : X \to Y$, $n = 1, 2, \ldots$, such that $f_n \to f$ pointwise, that is, $f_n(x) \to f(x)$ as $n \to \infty$ for each $x \in X$. If (M, d) is a metric space, $t \in M$ and $A \subset M$, we define

$$\mathrm{dist}(A, t) = \inf\{d(a, t) : t \in A\}.$$

Lemma 8.1.1.

Let M be a metric space, $(Z, \| \cdot \|)$ be a Banach space, $F : M \to 2^Z$ be a multivalued mapping, $B \subset Z$ be a closed ball centered at the origin, and $\epsilon > 0$. Assume that for every nonempty closed set $C \subset M$ there exist an open set

$U \subset M$, with $C \cap U \neq \emptyset$, and $z \in B$ such that dist $(F(t), z) < \epsilon$ for all $t \in C \cap U$. Then there exists a Baire class 1 mapping $f : M \to (Z, \|\cdot\|)$ such that $f(M) \subset B$ and

$$\mathrm{dist}(F(t), f(t)) < \epsilon \quad \text{for each} \quad t \in M.$$

Proof. For an open set $U \subset M$ and for $n = 1, 2, \ldots$ we consider functions $h_{U,n} : M \to [0, 1]$ defined by

$$h_{U,n}(t) = \min\big(1, n \cdot \mathrm{dist}(M \backslash U, t)\big), \qquad t \in M.$$

Let χ_U be the characteristic function of U, that is, $\chi_U(t) = 1$ if $t \in U$ and $\chi_U(t) = 0$ if $t \in M \backslash U$. We observe that the functions $h_{U,n}$ are continuous, $0 \le h_{U,n} \le \chi_U$ and $h_{U,n} \to \chi_U$ pointwise as $n \to \infty$. This shows that χ_U is a Baire class 1 function whenever U is an open set in M.

Consider the family $\mathcal{U}$ of all open sets $U \subset M$ such that there is a Baire class 1 mapping $f_U : M \to Z$ satisfying $f_U(M) \subset B$ and dist$\big(F(t), f_U(t)\big) < \epsilon$ for all $t \in U$. Then we put $U_0 = \bigcup \{U : U \in \mathcal{U}\}$. We shall show that $U_0 \in \mathcal{U}$. If $U_0 = \emptyset$, we are done. Further assume that $U_0 \neq \emptyset$; then $\mathcal{U} \neq \emptyset$. Recall that metric spaces are paracompact [En, Theorem 5.1.3]. Thus there exists a cover $\mathcal{W}$ of U_0, consisting of open sets such that (i) for every $W \in \mathcal{W}$ there is $U \in \mathcal{U}$ satisfying $W \subset U$ and (ii) for every $t \in U_0$ there is a neighborhood Ω of t such that $\Omega \cap W \neq \emptyset$ for only finitely many $W \in \mathcal{W}$. Observing that $W \in \mathcal{U}$ whenever $W \in \mathcal{W}$ and $W \subset U \in \mathcal{U}$, we can see that $\mathcal{W} \subset \mathcal{U}$. Let us well order $\mathcal{W}$, that is, write $\mathcal{W} = \{U_\gamma : \gamma \in [1, \xi)\}$, where ξ is an ordinal number [En, Section I.4]. For $\gamma \in [1, \xi)$ and $n = 1, 2, \ldots$ we put

$$a_{\gamma,n} = h_{U_\gamma,n}\left(1 - h_{\bigcup_{\beta<\gamma} U_\beta, n}\right).$$

Then $0 \le a_{\gamma,n} \le \chi_{U_\gamma}$ and $a_{\gamma,n}$ converge pointwise to the characteristic function of $U_\gamma \backslash \bigcup_{\beta<\gamma} U_\beta$. Next, it follows from the definition of $\mathcal{U}$ that for each $\gamma \in [1, \xi)$ we may find continuous mappings $f_{\gamma,n} : M \to Z$, $n = 1, 2, \ldots$, which converge pointwise to f_{U_γ}. Also, for each $n \in \mathbb{N}$ we put

$$f_n(t) = \sum_{\gamma \in [1,\xi)} a_{\gamma,n}(t) f_{\gamma,n}(t), \qquad t \in M.$$

As the family $\{U_\gamma : \gamma \in [1, \xi)\}$ has property (ii), each f_n is well defined and continuous. Finally, define $f_{U_0} : M \to Z$ by

$$f_{U_0} = \sum_{\gamma \in [1,\xi)} \chi_{U_\gamma}\big(1 - \chi_{\bigcup_{\beta<\gamma} U_\beta}\big) f_{U_\gamma}.$$

We immediately get that $f_n \to f_{U_0}$ pointwise as $n \to \infty$; hence f_{U_0} is Baire class 1. Moreover

$$\text{dist}\Big(F(t), f_{U_0}(t)\Big) < \epsilon \quad \text{for all} \quad t \in U_0.$$

This shows that U_0 lies in $\mathcal{U}$.

The proof will be completed when we show that $U_0 = M$. Assume this is not the case. Then, by assumption, there exist an open set $U \subset M$, with $(M \backslash U_0) \cap U \neq \emptyset$, and an element $z \in B$ such that

$$\text{dist}(F(t), z) < \epsilon \quad \text{for all} \quad t \in U \backslash U_0.$$

Then the mapping $f_U = \chi_{U \backslash U_0} z + \chi_{U_0} f_{U_0}$ is Baire class 1, $f_U(M) \subset B$, and dist $(F(t), f_U(t)) < \epsilon$ for all $t \in U$, which contradicts the definition of U_0 and the fact that the set $U \backslash U_0$ was nonempty. $\qquad \square$

Let X, Y be two topological spaces. We recall that a multivalued mapping $F : X \to 2^Y$ is said to be *upper semicontinuous* if the set

$$F^{-1}(C) := \{x \in X : F(x) \cap C \neq \emptyset\}$$

is closed for each closed set C in Y. The mapping F is said to be *usco* if it is upper semicontinuous and Fx is a nonempty compact set for each $x \in X$. It is called *minimal usco* if $F = G$ whenever $G : X \to 2^Y$ is a usco mapping such that $Gx \subset Fx$ for all $x \in X$.

Theorem 8.1.2.

Let M be a complete metric space, $(V, \| \cdot \|)$ be an Asplund space, and $G : M \to 2^{(B_{V^*}, w^*)}$ a usco mapping. Then there exists a Baire class 1 mapping $g : M \to (B_{V^*}, \| \cdot \|)$ such that $g(t) \in G(t)$ for all $t \in M$.

Proof. Consider a sequence of positive numbers $\epsilon_0, \epsilon_1, \dots$ such that $\epsilon_0 + \epsilon_1 + \cdots < +\infty$. Put $Z = V^*$, $F = G$, $B = B_{V^*}$, and $\epsilon = \epsilon_0$ in Lemma 8.1.1; we shall verify its assumptions. So let $\emptyset \neq C \subset M$ be any closed set. Let $H : C \to 2^{(B_{V^*}, w^*)}$ be a minimal usco mapping such that $H(t) \subset G(t)$ for every $t \in C$; such a mapping exists by Zorn's lemma. Since $H(C) \subset B_{V^*}$ and V is Asplund, Theorem 1.1.1 yields a weak* open set $W \subset V^*$ such that the set $H(C) \cap W$ is nonempty and has (norm) diameter less than ϵ_0. Using Lemma 3.1.2 and the minimality of H, we find an open set $U \subset M$ such that $C \cap U \neq \emptyset$ and $H(C \cap U) \subset W$; then the diameter of $H(C \cap U)$ is less than ϵ_0. Pick some $v^* \in H(C \cap U)$; so $v^* \in B_{V^*}$. Thus

$$\text{dist}(G(t), v^*) \leq \text{dist}(H(t), v^*) < \epsilon_0 \quad \text{for every} \quad t \in C \cap U.$$

Now we can apply Lemma 8.1.1. Hence there exists a Baire class 1 mapping $g_0 : M \to (V^*, \| \cdot \|)$ such that $g_0(M) \subset B_{V^*}$ and

$$\text{dist}(G(t), g_0(t)) < \epsilon_0 \quad \text{for every} \quad t \in M.$$

Next take $Z = V^*$, $B = (\epsilon_0 + \epsilon_1)B_{V^*}$, $F = G - g_0$, $\epsilon = \epsilon_1$. We shall again verify applicability of Lemma 8.1.1. Let $\emptyset \neq C \subset M$ be any closed set. Since g_0 is Baire class 1, there are $t_0 \in C$ and an open set $t_0 \in U_1 \subset M$ such that

$$\|g_0(t) - g_0(t_0)\| < \frac{\epsilon_1}{2} \quad \text{for every} \quad t \in C \cap U_1$$

[DGZ2, Theorem I.4.1; Ku, §31.X, Theorem 1]. (This is where the completeness of M is used.) Thus, since $G(t) \cap \left(g_0(t) + \epsilon_0 B_{V^*}\right) \neq \emptyset$ for every $t \in M$, we get that

$$D(t) := G(t) \cap \left[g_0(t_0) + (\epsilon_0 + \epsilon_1/2)B_{V^*}\right] \neq \emptyset \quad \text{for every} \quad t \in C \cap U_1.$$

Note that D is a usco mapping from $C \cap U_1$ to (B_{V^*}, w^*); see Lemma 3.1.1. Hence there exists a minimal usco mapping $H : C \cap U_1 \to 2^{(B_{V^*}, w^*)}$ such that $H(t) \subset D(t)$ for every $t \in C \cap U_1$. Since $H(C \cap U_1)$ is bounded and V is Asplund, we may find, as in the previous paragraph, an open set $U_2 \subset M$ such that $C \cap U_1 \cap U_2 \neq \emptyset$ and $H(C \cap U_1 \cap U_2)$ has diameter less than $\epsilon_1/2$. Put $U = U_1 \cap U_2$ and pick $v^* \in H(C \cap U)$. Then for every $t \in C \cap U$ we have

$$\text{dist}(G(t), v^*) \leq \text{dist}(H(t), v^*) < \frac{\epsilon_1}{2}$$

and

$$\text{dist}\left(G(t) - g_0(t), v^* - g_0(t_0)\right) < \frac{\epsilon_1}{2} + \frac{\epsilon_1}{2} = \epsilon_1.$$

Moreover, $v^* - g_0(t_0) \in (\epsilon_0 + \epsilon_1/2)B_{V^*} (\subset B)$. We have thus verified the hypotheses of Lemma 8.1.1. So there exists a Baire class 1 mapping $g_1 : M \to (V^*, \|\cdot\|)$ such that $g_1(M) \subset (\epsilon_0 + \epsilon_1)B_{V^*}$ and

$$\text{dist}\left(G(t), g_0(t) + g_1(t)\right) = \text{dist}\left(G(t) - g_0(t), g_1(t)\right) < \epsilon_1 \quad \text{for every} \quad t \in M.$$

Next, by repeating the above procedure, we can find a Baire class 1 mapping $g_2 : M \to (V^*, \|\cdot\|)$ such that $g_2(M) \subset (\epsilon_1 + \epsilon_2)B_{V^*}$ and

$$\text{dist}\left(G(t), g_0(t) + g_1(t) + g_2(t)\right) < \epsilon_2 \quad \text{for every} \quad t \in M,$$

and so on.

In this way we obtain for every $n = 1, 2, \ldots$ a Baire class 1 mapping $g_n : M \to (V^*, \|\cdot\|)$ such that $g_n(M) \subset (\epsilon_{n-1} + \epsilon_n)B_{V^*}$ and

$$\text{dist}\left(G(t), g_0(t) + \cdots + g_n(t)\right) < \epsilon_n \quad \text{for every} \quad t \in M.$$

Denote

$$g(t) = \sum_{n=0}^{\infty} g_n(t), \qquad t \in M.$$

This sum converges uniformly since $\epsilon_0 + \epsilon_1 + \cdots < +\infty$. So g is well defined and is Baire class 1 by [Č, Theorem 14.2.1]. Moreover, for every $t \in M$ we have

$$\text{dist}(G(t), g(t)) < \epsilon_n + \sum_{i=n+1}^{\infty} \|g_i(t)\| < \epsilon_n + \sum_{i=n}^{\infty} (\epsilon_i + \epsilon_{i+1}) \to 0$$

as $n \to \infty$. So $g(t) \in \overline{G(t)}$, and as $G(t)$ is weak* compact, $g(t) \in G(t)$ for all $t \in M$. Finally, let $f_n : M \to (V^*, \|\cdot\|)$ be continuous mappings such that $\|f_n(t) - g(t)\| \to 0$ as $n \to \infty$ for all $t \in M$. Put

$$h_n(t) = \frac{f_n(t)}{\max(\|f_n(t)\|, 1)}, \qquad t \in M.$$

Then h_n are continuous, $h_n(M) \subset B_{V^*}$ and $\|h_n(t) - g(t)\| \to 0$ as $n \to \infty$. Therefore g is a Baire class 1 mapping from M to $(B_{V^*}, \|\cdot\|)$. □

Lemma 8.1.3.

Consider a nonempty set Γ and a bounded sequence $\{g_k\}$ in $l_\infty(\Gamma)$ and let Δ be a subset of Γ such that whenever $\lambda_1, \lambda_2, \ldots \geq 0$ and $\lambda_1 + \lambda_2 + \cdots < +\infty$, then there exists $\gamma \in \Delta$ satisfying $\|\lambda_1 g_1 + \lambda_2 g_2 + \cdots\| = \lambda_1 g_1(\gamma) + \lambda_2 g_2(\gamma) + \cdots$. Then

$$\sup\left\{\limsup_{k \to \infty} g_k(\gamma) : \gamma \in \Delta\right\} \geq \inf\left\{\|g\| : g \in \text{co}\{g_k\}\right\}.$$

Proof. Denote

$$A = \inf\{\|g\| : g \in \text{co}\{g_k\}\}, \qquad B = \sup\{\|g_k\| : k = 1, 2, \ldots\}.$$

Then $0 \leq A \leq B < +\infty$. Let $\delta > 0$ be arbitrary and choose $\lambda \in (0, 1)$ such that

$$A - \delta(1 + \lambda) - B\lambda > (A - 2\delta)(1 - \lambda).$$

For $m = 1, 2, \ldots$ we choose inductively $h_m \in \text{co}\{g_k : k \geq m\}$ so that

$$\left\|\sum_{k=1}^{m} \lambda^{k-1} h_k\right\| < \inf\left\{\left\|\sum_{k=1}^{m-1} \lambda^{k-1} h_k + \lambda^{m-1} h\right\| : h \in \text{co}\{g_k : k \geq m\}\right\} + \delta\left(\frac{\lambda}{2}\right)^m.$$

(Here we put $\sum_{k=1}^{0} \ldots = 0$.) Since

$$\frac{h_m + \lambda h_{m+1}}{1 + \lambda} \in \text{co}\{g_k : k \geq m\} \quad \text{for all} \quad m \in \text{IN},$$

we have

$$\left\| \sum_{k=1}^{m} \lambda^{k-1} h_k \right\| < \left\| \sum_{k=1}^{m-1} \lambda^{k-1} h_k + \lambda^{m-1} \frac{h_m + \lambda h_{m+1}}{1 + \lambda} \right\| + \delta \left(\frac{\lambda}{2} \right)^m.$$

We now put $f_0 \equiv 0$, $f_m = \sum_{k=1}^{m} \lambda^{k-1} h_k$, for $m = 1, 2, \ldots$, and $f = \sum_{k=1}^{\infty} \lambda^{k-1} h_k$. Note that f is well defined as $0 < \lambda < 1$ and $\{h_k\}$ is a bounded sequence. Then, multiplying the above inequality by $(1 + \lambda)$, we get, for $m = 1, 2, \ldots$,

$$(1 + \lambda)\|f_m\| < \|\lambda f_{m-1} + f_{m+1}\| + \delta(1 + \lambda)\left(\frac{\lambda}{2}\right)^m$$

$$\leq \lambda \|f_{m-1}\| + \|f_{m+1}\| + \delta(1 + \lambda)\left(\frac{\lambda}{2}\right)^m,$$

and consequently,

$$\frac{\|f_{m+1}\| - \|f_m\|}{\lambda^m} > \frac{\|f_m\| - \|f_{m-1}\|}{\lambda^{m-1}} - \frac{\delta(1 + \lambda)}{2^m}.$$

Since $\|f_1\| - \|f_0\| = \|f_1\| \geq A$, we get from the last inequality that, for all $m \in \mathrm{IN}$,

$$\frac{\|f_m\| - \|f_{m-1}\|}{\lambda^{m-1}} > A - \delta(1 + \lambda)\left(\tfrac{1}{2} + \tfrac{1}{4} + \cdots\right) = A - \delta(1 + \lambda).$$

Hence

$$\|f\| - \|f_{m-1}\| = \sum_{k=m}^{\infty} \left(\|f_k\| - \|f_{k-1}\| \right) > \sum_{k=m}^{\infty} \lambda^{k-1} \left(A - \delta(1 + \lambda) \right),$$

that is,

$$\|f\| - \|f_{m-1}\| > \frac{\lambda^{m-1}}{1 - \lambda} \left(A - \delta(1 + \lambda) \right).$$

Now, by the assumption, there is $\gamma \in \Delta$ such that $f(\gamma) = \|f\|$. Then, for $m = 1, 2, \ldots$, we have

$$\lambda^{m-1} h_m(\gamma) = f(\gamma) - f_{m-1}(\gamma) - \sum_{k=m+1}^{\infty} \lambda^{k-1} h_k(\gamma)$$

$$\geq \|f\| - \|f_{m-1}\| - \sum_{k=m+1}^{\infty} \lambda^{k-1} B$$

$$> \frac{\lambda^{m-1}}{1 - \lambda} \left(A - \delta(1 + \lambda) \right) - \frac{\lambda^m}{1 - \lambda} B.$$

Hence, from the choice of λ, we get that $h_m(\gamma) \geq A - 2\delta$. Take any $m \in \mathrm{IN}$. Since h_m belongs to $\mathrm{co}\{g_k : k \geq m\}$, there is $k(m) \geq m$ such that $g_{k(m)}(\gamma) \geq A - 2\delta$. Thus we obtain

$$\limsup_{k \to \infty} g_k(\gamma) \geq A - 2\delta.$$

Now, the conclusion of our lemma follows since $\delta > 0$ was arbitrary.   □

8.2. P.R.I. IN DUALS OF ASPLUND SPACES

The aim of this section is to construct, in the dual of a general Asplund space, a P.R.I.; see Definition 6.1.5. By Proposition 6.1.7 we know that a P.R.I. can be constructed with the help of a projectional generator. Let us recall that a projectional generator on a Banach space Y is a couple (W, Φ) where W is a one-norming subset of Y^*, with $\overline{W}$ linear, and Φ is a multivalued mapping assigning to each $w \in W$ a nonempty, at most countable set $\Phi(w) \subset Y$ and is such that $\Phi(B)^\perp \cap \overline{B}^* = \{0\}$ for every $\emptyset \neq B \subset W$ for which $\overline{B}$ is linear. So we shall be done when we construct a projectional generator on the dual of an arbitrary Asplund space.

Let us first consider a special case when the Banach space V has an equivalent Fréchet smooth norm $\| \cdot \|$. Then V is an Asplund space; see Section 1.7. We shall show how to construct a projectional generator on V. Put $W = V$. (We always assume that V is a subspace of V^{**}.) Let $\Phi(0) = 0$ and for $0 \neq v \in V$ let $\Phi(v)$ be the Fréchet derivative of the norm $\| \cdot \|$ at v. Then Φ is norm continuous at all points off the origin [Ph, Proposition 2.8].

Let now $\emptyset \neq B \subset V$ be a set such that $\overline{B}$ is linear. Then $\Phi(\overline{B})$ is a subset of $\overline{\Phi(B)}$ (Here the Fréchet differentiability of $\| \cdot \|$ is used.) and hence $\Phi(B)^\perp = \Phi(\overline{B})^\perp$; thus it suffices to show that $\Phi(\overline{B})^\perp \cap \overline{B}^*$ consists of the origin only. According to Remark 6.1.8, it is enough to prove that $\Phi(\overline{B})^\perp \cap \overline{B \cap B_V}^* = \{0\}$. So take v^{**} in this intersection. Then there is a net $\{b_\alpha\} \subset B \cap B_V$ converging weak* to v^{**}. Hence, in particular,

$$\langle \Phi(b), b_\alpha \rangle \to \langle v^{**}, \Phi(b) \rangle = 0 \quad \text{for every} \quad b \in \overline{B}.$$

But, by the Bishop–Phelps theorem [Ph, Theorem 3.19], the closure of the set $\{\Phi(b)_{|\overline{B}} : b \in \overline{B}\}$ contains the dual unit sphere of the subspace $\overline{B}$. Hence $b_\alpha \to 0$ weakly in the subspace $\overline{B}$ and so $b_\alpha \to 0$ weakly in the whole V. It follows that v^{**}, as a weak* limit of $\{b_\alpha\}$, must be equal to 0. We have thus verified that (V, Φ) is a projectional generator on V^*.

The case of a general Asplund space requires much more work. Difficulties arise in constructing the couple (W, Φ) as well as in proving that it is in fact a projectional generator. The new ingredients that we shall need are Theorem 8.1.2, a separable reduction, and Lemma 8.1.3.

Proposition 8.2.1.

A Banach space V is Asplund (if and) only if V^* admits a projectional generator (W, Φ) such that $W \subset V$ and $\overline{W} = V$; we can even take $W = V$.

Proof. As to the sufficiency, see the text below the proof of Proposition 6.1.9. Assume now $(V, \|\cdot\|)$ is Asplund. The subdifferential $\partial\|\cdot\|$ is easily checked to be a multivalued usco mapping from V into (B_{V^*}, w^*); see [Ph, Proposition 2.5]. Hence, according to Theorem 8.1.2, there is a Baire class 1 mapping $f_0 : (V, \|\cdot\|) \to (B_{V^*}, \|\cdot\|)$ such that $f_0(v) \in \partial\|\cdot\|(v)$ for every $v \in V$. Let $f_n : V \to B_{V^*}$, $n = 1, 2, \ldots$, be norm-to-norm continuous mappings such that

$$\|f_n(v) - f_0(v)\| \to 0 \quad \text{as} \quad n \to \infty \quad \text{for every} \quad v \in V.$$

Put

$$\Phi(v) = \{f_1(v), f_2(v), \ldots\}, \qquad v \in V.$$

Clearly, Φ is an at most countable-valued mapping from V to B_{V^*}. We shall prove that the couple (V, Φ) is a projectional generator on the space V^*.

So fix any nonempty set B in V with $\bar{B}$ linear. We have to show that

$$\Phi(B)^{\perp} \cap \overline{B \cap B_V}^* = \{0\}; \tag{1}$$

see Remark 6.1.8. Take any v^{**} in $\Phi(B)^{\perp} \cap \overline{B \cap B_V}^*$. Assume $v^{**} \neq 0$; then there is $v_0^* \in V^*$ such that $\langle v^{**}, v_0^* \rangle \neq 0$. We shall construct a sequence $\{v_n\} \subset B \cap B_V$ and a sequence of finite subsets $F_1 \subset F_2 \subset \ldots \subset V^*$ as follows. Fix any $0 \neq v_1 \in B \cap B_V$ and put $F_1 = \{v_0^*\}$. Suppose we have chosen $v_1, v_2, \ldots, v_n$ and $F_1, F_2, \ldots, F_n$ for some $n \geq 1$. Let H be a $1/n$-net in the set

$$\bigcup_{i=1}^{n} f_i\big(\mathrm{sp}\{v_1, \ldots, v_n\} \cap nB_V\big)$$

endowed with the metric generated by the dual norm. Note that, owing to the continuity of the f_i, the last set is compact; so we may and do assume that H is finite. Put now $F_{n+1} = F_n \cup H$ and find $v_{n+1} \in B \cap B_V$ such that

$$\big|\langle v^{**} - v_{n+1}, v^* \rangle\big| < \frac{1}{n} \quad \text{for each} \quad v^* \in F_{n+1}.$$

Performing this for each $n = 1, 2, \ldots$, we finally put $F = F_1 \cup F_2 \cup \cdots$ and $B_0 = \mathrm{sp}\{v_1, v_2, \ldots\}$. It is easy to check that F is (norm) dense in $\Phi(B_0)$ and that

$$\langle v^*, v_n \rangle \to \langle v^{**}, v^* \rangle \quad \text{for each} \quad v^* \in \Phi(B_0) \cup \{v_0^*\}.$$

Now $B_0 \subset \mathrm{sp}\, B \subset \mathrm{sp}\, \bar{B} = \bar{B}$, and by the continuity of f_i's, we have $\Phi(\bar{B}) \subset \overline{\Phi(B)}$. So $v^{**} \in \Phi(B)^{\perp}$ implies $v^{**} \in \Phi(B_0)^{\perp}$. Therefore $\langle v^{**}, v^* \rangle = 0$ and so

$$\langle v^*, v_n \rangle \to 0$$

for each v^* from $\Phi(B_0)$ and in fact for each v^* from $\overline{\mathrm{sp}}\,\Phi(B_0)$ since $\{v_n\}$ is bounded. Denote $Y = \bar{B}_0$; this will be a separable Asplund space.

It remains to show that

$$\{v^*|_Y :\ v^* \in \overline{\mathrm{sp}}\,\Phi(B_0)\} = Y^*. \tag{2}$$

For then $v_n \to 0$ weakly in Y, hence weakly in V, and thus $\langle v^{**}, v_0^*\rangle = \lim_n \langle v_0^*, v_n\rangle = 0$, a contradiction. Thus (1) will be proved. Suppose, for the purpose of obtaining a contradiction, that (2) is false. Then there exist $y_0^{**} \in Y^{**}$ and $y_0^* \in B_{Y^*}$ such that

$$\langle y_0^{**}, y_0^*\rangle > 0 = \langle y_0^{**}, v^*|_Y\rangle \quad \text{for all} \quad v^* \in \overline{\Phi(B_0)}.$$

Since Y is separable Asplund, Y^* is separable; hence, by Goldstine's theorem, there is a bounded sequence $\{y_k\}$ in Y which converges weak* to y_0^{**}. (Yes, sequences are enough since $(B_{Y^{**}}, w^*)$ is metrizable.) Without loss of generality we may assume that

$$\langle y_0^*, y_k\rangle > \tfrac{1}{2}\langle y_0^{**}, y_0^*\rangle \quad \text{for all} \quad k = 1, 2, \dots.$$

Now we are ready to apply Lemma 8.1.3. Put there $\Gamma = B_{Y^*}$, $\Delta = \{v^*|_Y :\ v^* \in \overline{\Phi(B_0)}\}$, and $g_k = y_k$, $k = 1, 2, \dots$. From the definition of Φ, we can deduce that the premise of the lemma is satisfied. Thus

$$\begin{aligned}
0 &= \sup\Big\{\langle y_0^{**}, v^*|_Y\rangle :\ v^* \in \overline{\Phi(B_0)}\Big\}\\
&= \sup\Big\{\lim_{k\to\infty} \langle v^*|_Y, y_k\rangle :\ v^* \in \overline{\Phi(B_0)}\Big\}\\
&\geq \inf\Big\{\|y\| :\ y \in \mathrm{co}\{y_k\}\Big\}\\
&\geq \inf\Big\{\langle y_0^*, y\rangle :\ y \in \mathrm{co}\{y_k\}\Big\}\\
&> \tfrac{1}{2}\langle y_0^{**}, y_0^*\rangle > 0,
\end{aligned}$$

a contradiction. This proves (2) and finishes the whole proof. $\qquad\square$

Theorem 8.2.2.

Let V be a nonseparable Asplund space. Then V^* admits
   (i) a P.R.I. $\{P_\alpha :\ \omega \leq \alpha \leq \mu\}$ such that for each $\omega \leq \alpha < \mu$ the space $(P_{\alpha+1} - P_\alpha)V^*$ is isometric to the dual of an Asplund space,
   (ii) a S.P.R.I.,
   (iii) a linear and continuous injection into $c_0(\Gamma)$ for some set Γ, and
   (iv) a Markuševič basis.

Proof. By putting together Propositions 8.2.1 and 6.1.9 and Theorem 1.1.2 (i), (ii), we get (i). Further, we proceed, by transfinite induction over dens V, as we did in the proof of Theorem 7.2.2. $\qquad\square$

8.3. ASPLUNDNESS AND WCG SPACES

Let us recall that, for a nonempty set Γ, the symbol $\Sigma(\Gamma)$ means the space of all $x \in \mathbb{R}^\Gamma$ that have a countable support. The topology on $\Sigma(\Gamma)$ is that inherited from the product topology of $\mathbb{R}^\Gamma$, that is, the topology p of pointwise convergence on Γ. Compacta that can be found, up to a homeomorphism, in $(\Sigma(\Gamma), p)$ for some set Γ are called *Corson*.

Proposition 8.3.1.

Let V be a Banach space such that its dual unit ball endowed with the weak* topology is Corson compact. Then V admits a projectional generator (W, Φ) with $W = V^*$.

Proof. We find a set Γ and a continuous injective mapping $S : (B_{V^*}, w^*) \to (\Sigma(\Gamma), p)$. For $v^* \in B_{V^*}$ we write $S(v^*) = \{S_\gamma(v^*) : \gamma \in \Gamma\}$. Then S_γ are real-valued continuous functions on (B_{V^*}, w^*). Let us observe that V can be canonically embedded into $C((B_{V^*}, w^*))$; denote this embedding by i. Since $i(V)$ separates the points of B_{V^*}, we know by the Stone–Weierstrass theorem that each S_γ can be approximated, in the supremum norm, by the algebra generated by $i(V)$ and the constant functions. For each $\gamma \in \Gamma$ let V_γ be a countable subset of V such that the function S_γ belongs to the closure of the algebra generated by $i(V_\gamma)$ and the constant functions. Now we are ready to define the mapping $\Phi : V^* \to 2^V$ by $\Phi(0) = 0$, $\Phi(\xi) = 0$ if $\|\xi\| > 1$ and

$$\Phi(\xi) = \bigcup \{V_\gamma : \gamma \in \Gamma \text{ satisfies } S_\gamma(\xi) \neq S_\gamma(0)\}, \quad \text{if} \quad 0 \neq \xi \in B_{V^*}.$$

Note that the sets $\Phi(\xi)$ are at most countable since $S(B_{V^*}) \subset \Sigma(\Gamma)$.

We shall prove that the couple (V^*, Φ) is a projectional generator on V. So let B be any nonempty subset of V^* such that $\overline{B}$ is linear. According to Remark 6.1.8, it is enough to show that $\Phi(B)^\perp \cap \overline{B \cap B_{V^*}}^*$ consists only of 0. So take any ξ in this intersection and assume, for the purpose of obtaining a contradiction, that $\xi \neq 0$. The injectivity of S says that $S(\xi) \neq S(0)$; so there must exist $\gamma \in \Gamma$ such that $S_\gamma(\xi) \neq S_\gamma(0)$. Since $\xi \in \overline{B \cap B_{V^*}}^*$ and S_γ is weak* continuous, there is $b \in B \cap B_{V^*}$ such that $S_\gamma(b) \neq S_\gamma(0)$. Then V_γ is a subset of $\Phi(b)$; see the definition of $\Phi(b)$. And since $\xi \in \Phi(B)^\perp$, we get that $i(v)(\xi) = \langle \xi, v \rangle = 0$ for all $v \in V_\gamma$. Also $i(v)(0) = 0$ for all $v \in V_\gamma$. Now, we know that S_γ lies in the closure of the algebra generated by $i(V_\gamma)$ and the constants. Hence $S_\gamma(\xi) = S_\gamma(0)$, a contradiction. It follows therefore that $\Phi(B)^\perp \cap \overline{B \cap B_{V^*}}^* = \{0\}$, which means that (V^*, Φ) is a projectional generator on V. $\square$

Proposition 8.3.2.

Let V be a Banach space such that (B_{V^*}, w^*) is a Corson compact. Let a subspace Y of V be Asplund. Then Y is WCG and admits a shrinking Markuševič basis; see Definition 6.2.3.

Proof. If Y is separable, then it is surely WCG and it has a shrinking Markuševič basis according to [LT, Proposition 1.f.4]. Let us assume that the proposition is verified whenever the density of Y is less than a given uncountable cardinal $\aleph$. Assume now that Y has density equal to $\aleph$. By putting together Propositions 8.2.1, 8.3.1, and 6.1.10, we get that Y admits a P.R.I. $\{P_\alpha : \omega \le \alpha \le \mu\}$ such that $\{P_\alpha^* : \omega \le \alpha \le \mu\}$ is a P.R.I. on Y^*.

Fix any $\omega \le \alpha < \mu$. Then the space $(P_{\alpha+1} - P_\alpha)Y$ is Asplund, has density less than $\aleph$, and is a subspace of V. Thus, by our induction assumption, the space $(P_{\alpha+1} - P_\alpha)Y$ is WCG and admits a shrinking Markuševič basis. Now, applying Proposition 6.2.5, we can conclude that the whole Y is WCG and has a shrinking Markuševič basis.   □

Theorem 8.3.3.

Let V be an Asplund space. Then the following assertions are equivalent:
  (i) V is WCG;
  (ii) V is a subspace of a WCG space;
  (iii) V is weakly $\mathcal{K}$-analytic, (see Definition 4.1.1);
  (iv) V is Vašák, (see Definition 7.1.5);
  (v) V is WLD, (see Definition 7.2.6);
  (vi) (B_{V^*}, w^*) is a Corson compact; and
  (vii) V has a shrinking Markuševič basis.

Proof. Here (i) $\Rightarrow$ (ii) $\Rightarrow$ (iii) $\Rightarrow$ (iv) $\Rightarrow$ (v) $\Rightarrow$ (vi) always hold; see Proposition 7.1.6 and Theorem 7.2.7. That (vi) $\Rightarrow$ (vii) follows from Proposition 8.3.2. Finally, (vii) $\Rightarrow$(i) holds since if $\{(v_\gamma, v_\gamma^*) : \gamma \in \Gamma\}$ is a shrinking Markuševič basis on V, then $\{\|v_\gamma\|^{-1} v_\gamma : \gamma \in \Gamma\} \cup \{0\}$ is easily seen to be a weakly compact set whose linear span is dense in V.   □

Here it should be added that *a Banach space has a shrinking Markuševič basis (if and) only if it is Asplund and WCG;* see the text below Definition 6.2.3.

Using an interpolation technique, we can extend Theorem 8.3.3 to Asplund generated spaces, thus obtaining a further characterization of WCG spaces. Let us recall that a Banach space is said to be *Asplund generated* if it contains, as a dense subset, the continuous linear image of an Asplund space.

Theorem 8.3.4.

A Banach space V is WCG if and only if it is Asplund generated and one of the following conditions is satisfied:
  (i) V is a subspace of a WCG space;
  (ii) V is weakly $\mathcal{K}$-analytic;

(iii) V is Vašák;
(iv) V is WLD; and
(v) (B_{V^*}, w^*) is a Corson compact.

Proof. Let V be WCG. Then, according to Theorem 1.2.3, we know that there exists a reflexive space which embeds linearly and continuously into a dense subset of V; so V is Asplund generated.

That (i) $\Rightarrow$ (ii) $\Rightarrow$ (iii) $\Rightarrow$ (iv) $\Rightarrow$ (v) always holds.

Assume now that (B_{V^*}, w^*) is Corson compact and that V is Asplund generated. Then there exists an Asplund space E such that $E \subset V$, $\overline{E} = V$, and $B_E \subset B_V$. We apply Theorem 1.4.4 to our V and to $M = B_E$. We get an Asplund space Y and a continuous linear mapping $T: Y \to V$ such that $\overline{TY} = V$. We recall that $Y = \{\{v_n\} \in Z : v_1 = v_2 = \cdots\}$, where $Z = \{\{v_n\} \in V^{\mathbb{N}} : \sum_{n=1}^{\infty} \|v_n\|_n^2 < +\infty\}$ and the norm on Z is defined by $\|\{v_n\}\| = (\sum_{n=1}^{\infty} \|v_n\|_n^2)^{1/2}$, where each $\|\cdot\|_n$ is some equivalent norm on V. It is easy to check that (B_{Z^*}, w^*) continuously injects into $\prod_{n=1}^{\infty} (B_{(V^*, \|\cdot\|_n)}, w^*)$ and that this product continuously injects into $(\Sigma(\Delta), p)$ for some Δ, because (B_{V^*}, w^*) is Corson compact. Hence (B_{Z^*}, w^*) is Corson compact too. Now, Proposition 8.3.2 applies and therefore Y must be WCG. Finaly, V is WCG since $\overline{TY} = V$. $\square$

The concepts used in the next theorem were introduced in Definitions 1.2.1, 1.5.1, and 7.1.7.

Theorem 8.3.5.

For a compact space K the following assertions are equivalent:
(i) K is Eberlein compact;
(ii) K is simultaneously Radon–Nikodým and Gul'ko compact; and
(iii) K is simultaneously Radon–Nikodým and Corson compact.

Proof. For (i)$\Rightarrow$(ii) see Propositions 1.5.2 and 7.2.3. That (ii)$\Rightarrow$(iii) follows from Theorem 7.2.7.

Suppose (iii) is satisfied. We shall prove (i). According to Theorem 1.2.4, we have to show that $C(K)$ is WCG. From Theorem 1.5.4 we know that $C(K)$ is Asplund generated. So, in order to apply Theorem 8.3.4, it remains to show that $C(K)$ is WLD. Thus, regarding Lemma 7.3.5, it is enough to prove that our K has the property that for every positive, regular Borel measure μ on K there exists a separable Borel set $B \subset K$ such that $\mu(B) = \mu(K)$.

Since K is Radon–Nikodým compact, we may and do assume that K is a subspace of (B_{Y^*}, w^*), where Y is some Asplund space. Take any positive regular Borel measure μ on K. We first claim that *for every $\epsilon > 0$ and for every Borel set $M \subset K$, with $\mu(M) > 0$, there is a closed set $L \subset M$ such that $\mu(L) > 0$*

and the norm-diameter diam L *of* L *is less than* ϵ. To prove this, fix such ϵ and M. Then there exists a closed set $L_1 \subset M$ such that $\mu(L_1) > 0$. Let U_0 be the union of the family of all open sets $U \subset K$ such that $L_1 \cap U \neq \emptyset$ and $\mu(L_1 \cap U) = 0$. A simple compactness argument together with the regularity of μ yields that $\mu(L_1 \cap U_0) = 0$. We put $L_2 = L_1 \backslash U_0$; then surely $L_2 \neq \emptyset$. From the weak* dentability of Y^* there is an open set $\Omega \subset K$ with $L_2 \cap \Omega \neq \emptyset$ and with diam$(L_2 \cap \Omega) < \epsilon$; see Theorem 1.1.1. We note that $\Omega \backslash U_0 \neq \emptyset$, so $\mu(L_1 \cap \Omega) > 0$. Thus

$$\mu(L_2 \cap \Omega) = \mu(L_1 \cap \Omega) - \mu(L_1 \cap U_0 \cap \Omega) = \mu(L_1 \cap \Omega) > 0.$$

Finally it remains to find a closed set $L \subset L_2 \cap \Omega$ with $\mu(L) > 0$. Then diam $L < \epsilon$ and the claim is proved.

Fix an arbitrary $n \in \mathbb{IN}$. Let $\mathcal{F}_n$ be a maximal family of mutually disjoint closed subsets L of K such that $\mu(L) > 0$ and diam $L < 1/n$. Then, as $\mu(K) < +\infty$, for every $\epsilon > 0$ we have $\mu(L) > \epsilon$ only for finitely many L in $\mathcal{F}_n$. It follows that $\mathcal{F}_n$ is at most countable. We put $H_n = \bigcup \{L : L \in \mathcal{F}_n\}$; this set is F_σ, and hence Borel. Assume that $\mu(H_n) < \mu(K)$, so $\mu(K \backslash H_n) > 0$. Hence, by the claim from the previous paragraph, there is a closed set $L \subset K \backslash H_n$ with $\mu(L) > 0$ and diam $L < 1/n$, a contradiction with the maximality of $\mathcal{F}_n$. Therefore $\mu(H_n) = \mu(K)$. Next, since H_n is the union of at most countably many sets of diameter less than $1/n$, there is an at most countable set $S_n \subset H_n$ such that $H_n \subset S_n + (1/n)B_{Y^*}$.

Put $B = \bigcap_{n=1}^\infty H_n$ and $S = \bigcup_{n=1}^\infty S_n$. Note that B is a Borel set with $\mu(B) = \mu(K)$ and S is countable. Since

$$B \subset H_n \subset S_n + \frac{1}{n}B_{Y^*} \subset S + \frac{1}{n}B_{Y^*}$$

for all $n \in \mathbb{IN}$, the set B is norm separable. Hence B is weak* separable, which means that it is separable in the topology of the space K. Now we may apply Lemma 7.3.5 and finish the proof. $\square$

We can extend the above theorem as follows:

Theorem 8.3.6.

Let $\varphi : K \to L$ *be a continuous mapping from a Radon–Nikodým compact K onto a Corson compact L. Then φ factors through an Eberlein compact S, that is, there are continuous mappings $\alpha : K \to S$ and $\tilde{\varphi} : S \to L$ such that $\tilde{\varphi} \circ \alpha = \varphi$; hence L itself is an Eberlein compact.*

Proof. Since L is a Corson compact, $L \subset (\Sigma(\Gamma), p)$ for some set Γ. Using the mapping ψ from the proof of Lemma 7.3.5, we can see that L is homeomorphic with a compact lying in $\Sigma(\Delta) \cap [0, +\infty)^\Delta$ for some set Δ. Hence we may and do assume that $L \subset \Sigma(\Gamma) \cap [0, +\infty)^\Gamma$. For $\gamma \in \Gamma$ we define $\pi_\gamma \in C(L)$ by

$\pi_\gamma(l) = l_\gamma$, $l = \{l_{\gamma'}\} \in L$. We note that the family $\{\pi_\gamma : \gamma \in \Gamma\}$ separates the points of L, that is, if $l, l' \in L$ are distinct, then $\pi_\gamma(l) \neq \pi_\gamma(l')$ for some $\gamma \in \Gamma$, and is point countable, that is, the set $\{\gamma \in \Gamma : \pi_\gamma(l) \neq 0\}$ is at most countable for every $l \in L$.

Since K is Radon–Nikodým compact, $C(K)$ is an Asplund generated space according to Theorem 1.5.4. Hence there is a closed, convex, symmetric, and Asplund set $D \subset B_{C(K)}$ such that $\bigcup_{n=1}^\infty nD$ is dense in $C(K)$. By Lemma 1.4.3(i), we may and do assume that the constant function 1 belongs to D. For $m, n \in \mathrm{IN}$ we put $\Gamma_n^m = \{\gamma \in \Gamma : \mathrm{dist}(\pi_\gamma \circ \varphi, nD) < 1/m\}$. (Here $\mathrm{dist}(f, M) = \inf\{\|f - g\| : g \in M\}$.) Then for each $\gamma \in \Gamma_n^m$ we find $h_{n,\gamma}^m \in D$ such that $\|\pi_\gamma \circ \varphi - nh_{n,\gamma}^m\| < 1/m$. Clearly, $\bigcup_{n=1}^\infty \Gamma_n^m = \Gamma$ for each $m \in N$. Finally, we put

$$u_{n,\gamma}^m = \max\left(h_{n,\gamma}^m - \frac{1}{mn}, 0\right), \qquad m, n \in \mathrm{IN}, \qquad \gamma \in \Gamma_n^m;$$

note that $u_{n,\gamma}^m \in C(K)$.

We claim that *the family* $A := \{u_{n,\gamma}^m : \gamma \in \Gamma_n^m, m, n \in \mathrm{IN}\}$ *separates those* $k, k' \in K$ *for which* $\varphi(k) \neq \varphi(k')$. So fix such k, k'. We find $\gamma \in \Gamma$ such that $\pi_\gamma(\varphi(k)) \neq \pi_\gamma(\varphi(k'))$, say $\pi_\gamma(\varphi(k)) > \pi_\gamma(\varphi(k'))$. Then we may find $m \in \mathrm{IN}$ satisfying $\pi_\gamma(\varphi(k)) > \pi_\gamma(\varphi(k')) + 2/m$. Finally, we choose $n \in \mathrm{IN}$ such that $\gamma \in \Gamma_n^m$; then $\|\pi_\gamma \circ \varphi - nh_{n,\gamma}^m\| < 1/m$. Thus $h_{n,\gamma}^m(k) > h_{n,\gamma}^m(k')$ and

$$h_{n,\gamma}^m(k) - \frac{1}{mn} > \frac{1}{n}\pi_\gamma(\varphi(k)) - \frac{2}{mn} > \frac{1}{n}\pi_\gamma(\varphi(k')) \geq 0.$$

Therefore

$$u_{n,\gamma}^m(k') = \max\left(h_{n,\gamma}^m(k') - \frac{1}{mn}, 0\right) < h_{n,\gamma}^m(k) - \frac{1}{mn} = u_{n,\gamma}^m(k).$$

We further claim that *the family* A *is point countable*. To prove this, we fix $k \in K$ and $m, n \in \mathrm{IN}$. Assume that $u_{n,\gamma}^m(k) \neq 0$ for some $\gamma \in \Gamma_n^m$. Then we have $h_{n,\gamma}^m(k) > 1/(mn)$ and so $\pi_\gamma(\varphi(k)) > 0$. We recall that the family $\{\pi_\gamma : \gamma \in \Gamma\}$ is point countable. Therefore the set of all $\gamma \in \Gamma_n^m$ for which $u_{n,\gamma}^m(k) \neq 0$ is at most countable. Now we are done because m, n run throughout the countable set IN.

For $k \in K$ we put

$$[k] = \{k' \in K : u_{n,\gamma}^m(k') = u_{n,\gamma}^m(k) \quad \text{for all} \quad \gamma \in \Gamma_n^m \quad \text{and all} \quad m, n \in \mathrm{IN}\}$$

and denote $S = \{[k] : k \in K\}$. We define $\alpha : K \to S$ by $\alpha(k) = [k]$, $k \in K$ and $\widetilde{\varphi} : S \to L$ by $\widetilde{\varphi}([k]) = \varphi(k)$, $[k] \in S$. By the first claim, $\widetilde{\varphi}$ is well defined, and clearly $\widetilde{\varphi} \circ \alpha = \varphi$. We endow S with the topology: $U \subset S$ is open if and only if $\alpha^{-1}(U)$ is an open set. Then, clearly, α and $\widetilde{\varphi}$ will be continuous.

Now put $\Delta = \bigcup_{m,n=1}^\infty \Gamma_n^m \times \{m\} \times \{n\}$ and define the mappings $\Phi : K \to \mathrm{IR}^\Delta$ and $\widetilde{\Phi} : S \to \mathrm{IR}^\Delta$ by

$$\Phi(k)(\gamma, m, n) = u_{n,\gamma}^m(k), \qquad (\gamma, m, n) \in \Delta, \qquad k \in K,$$

$$\widetilde{\Phi}([k]) = \Phi(k), \qquad [k] \in S.$$

Clearly, Φ is continuous, and by the second claim, $\Phi(K) \subset \Sigma(\Delta)$. Further, it is straightforward to check that $\widetilde{\Phi}$ is well defined and injective and that $\widetilde{\Phi} \circ \alpha = \Phi$. Hence $\widetilde{\Phi}$ must be continuous and so S is shown to be Corson compact.

Finally, put $Y = \overline{\mathrm{sp}}\, A$ (recall that $Y \subset C(K)$) and we check that for each fixed $k \in K$ the assignment $y \mapsto y(k)$, $y \in Y$, is an element of B_{Y^*}. Next we define the mappings $\Psi : K \to (B_{Y^*}, w^*)$ and $\widetilde{\Psi} : S \to (B_{Y^*}, w^*)$ by

$$\Psi(k)(y) = y(k), \qquad k \in K, \qquad y \in Y,$$

$$\widetilde{\Psi}([k]) = \Psi(k), \qquad [k] \in S.$$

Then Ψ is continuous, $\widetilde{\Psi}$ is well defined and injective, $\widetilde{\Psi} \circ \alpha = \Psi$, and hence $\widetilde{\Psi}$ is continuous. Now we claim that A *is an Asplund set*. Assuming this claim is true, it follows from Theorem 1.4.4 that Y is an Asplund generated space. Therefore (B_{Y^*}, w^*) is Radon–Nikodým compact and so is S.

It remains to prove the claim. For $M \subset C(K)$, we put $M \vee 0 = \{\max(f, 0) : f \in M\}$. Thus

$$A \subset \bigcup_{m,n=1}^{\infty} \left(D - \frac{1}{mn} \right) \vee 0 \subset 2D \vee 0,$$

and it is enough to prove that $D \vee 0$ is an Asplund set. From the Weierstrass theorem [Ru2, Theorem 15.26], for $n = 1, 2, \ldots$, we may find polynomials p_n such that

$$|p_n(t) - \max(t, 0)| < \frac{1}{n} \quad \text{for all} \quad t \in [-1, 1].$$

Thus $\|p_n \circ f - \max(f, 0)\| < 1/n$ for all $f \in D$. But

$$p_n \circ f \in a_1 D + a_2 D^2 + \cdots + a_{j_n} D^{j_n} =: E_n$$

for some $a_1, \ldots, a_{j_n} \in \mathrm{IR}$ and some $j_n \in \mathrm{IN}$. We note that the sets E_n are Asplund according to Lemma 1.4.3(i) and to the argument from the last paragraph of the proof of Theorem 1.5.4. Therefore we have $D \vee 0 \subset \bigcap_{n=1}^{\infty} \left(E_n + (1/n)B_{C(K)} \right)$, and so Lemma 1.4.3(iii) guarantees that $D \vee 0$ is an Asplund set.

Summarizing, we have that S is both Corson compact and Radon–Nikodým compact. It then follows from Theorem 8.3.5 that S is Eberlein compact. Finally using a theorem of Benyamini, Rudin, and Wage [BRW], we get that L $(= \widetilde{\varphi}(S))$ is Eberlein compact as well. $\qquad \square$

Now the Banach space counterpart of the above theorem follows.

Theorem 8.3.7.

A Banach space V is isomorphic to a subspace of a WCG space if and only if it is isomorphic to a subspace of an Asplund generated space and one of the following conditions is satisfied:

    (i) V is weakly $\mathcal{K}$-analytic;

    (ii) V is Vašák;

    (iii) V is WLD; and

    (iv) (B_{V^*}, w^*) is a Corson compact.

Proof. Assume that V is isomorphic to a subspace of an Asplund generated space and let (iv) be satisfied. Then, Theorem 8.3.6 says that (B_{V^*}, w^*) is (a continuous image of an) Eberlein compact, K, say. Thus, V isometrically embeds into $C(K)$, which is WCG by Theorem 1.2.4.

The remaining implications are either trivial or have already been proved. $\qquad\qquad\square$

8.4. A WEAKLY $\mathcal{K}$-ANALYTIC SPACE NOT CONTAINING ℓ_1 MAY NOT BE A SUBSPACE OF A WCG SPACE

We note that if a Banach space is Asplund, then it does not contain ℓ_1 isomorphically. Indeed, this is so because $\ell_1^* = \ell_\infty$ and this space is not separable. This suggests a natural question: *Can we replace the Asplundness in Theorem 8.3.3 by the noncontainment of ℓ_1?* In what follows, we shall show that the implication (iii)$\Rightarrow$(ii) in Theorem 8.3.3 fails if we replace the Asplundness of V by the condition that V does not contain an isomorphic copy of ℓ_1. In building this counterexample, we shall put together two constructions: James's tree space [J] and Rezničenko's compact [K5, Arg].

REZNIČENKO'S COMPACT

Denote $\Gamma = \left(\{0\} \times \{1, \frac{1}{2}, \frac{1}{3}, \ldots\}\right) \cup \left((0, \omega_1) \times [0, 1]\right)$, where ω_1 means the first uncountable ordinal. We shall construct families $\mathcal{A}_\alpha$, $\alpha \in [0, \omega_1)$, of finite sequences $\{(\xi_1, t_1), \ldots, (\xi_k, t_k)\}$ in Γ such that $k \in \mathbb{N}$, $0 = \xi_1 < \xi_2 < \cdots < \xi_k < \omega_1$, and with the property $\#\mathcal{A}_\alpha \leq 2^{\aleph_0}$ as follows: Put $\mathcal{A}_0 = \{\{(0, 1)\}, \{(0, \frac{1}{2})\}, \{(0, \frac{1}{3})\}, \ldots\}$. Fix $0 < \alpha < \omega_1$ and assume we have already constructed the families $\mathcal{A}_\beta$ for all $0 \leq \beta < \alpha$. We observe that $\#\bigcup_{\beta<\alpha}\mathcal{A}_\beta \leq 2^{\aleph_0} \cdot \aleph_0 = 2^{\aleph_0}$. Denote by $\mathcal{M}$ the family of all countable subsets $\{M_1, M_2, \ldots\}$ of $\bigcup_{\beta<\alpha}\mathcal{A}_\beta$ such that $M_i \cap M_j = \emptyset$ whenever $i \neq j$. Then $\#\mathcal{M} \leq \#\left(\bigcup_{\beta<\alpha}\mathcal{A}_\beta\right)^{\mathbb{N}} \leq 2^{\aleph_0}$. Let $g : [0, 1] \to \mathcal{M}$ be a surjective mapping; it certainly exists since $\#[0, 1] = 2^{\aleph_0}$. We define

$$\mathcal{A}_\alpha = \{M \cup \{(\alpha, t)\} : \ M \in g(t)\}.$$

Finally, we put

$$A = \bigcup_{\alpha < \omega_1} \mathcal{A}_\alpha,$$

$$\mathcal{A}' = \Big\{ ([\xi, \omega_1) \times [0, 1]) \cap M : \ \xi \in [0, \omega_1), \ M \in \mathcal{A} \Big\}$$

and

$$K = \mathrm{cl}\{\chi_M : \ M \in \mathcal{A}'\},$$

where the closure is considered in the space $\{0, 1\}^\Gamma$. (Here χ_M means the characteristic function of the set M.) This K is called *Rezničenko's compact*, and a set $\mathcal{R} \subset 2^\Gamma$ satisfying

$$K = \{\chi_M : \ M \in \mathcal{R}\}$$

is called *Rezničenko's family*. If $M = \{(\xi_1, t_1), \ldots, (\xi_k, t_k)\} \in \mathcal{A}'$, then we put $\min M = (\xi_1, t_1)$ and $\max M = (\xi_k, t_k)$. If $(\xi, t) \in \Gamma$, then we define $\pi(\xi, t) = \xi$. In what follows we list some easily provable facts concerning the family $\mathcal{A}$:

(i) If $M, N \in \mathcal{A}$, $\{(\xi_1, t_1), (\xi_2, t_2)\} \subset M \cap N$, and $\xi_1 < \xi_2$, then $M \cap ([0, \xi_2] \times [0, 1]) = N \cap ([0, \xi_2] \times [0, 1]) \in \mathcal{A}$.

(ii) For every $\gamma \in \Gamma$, with $\pi(\gamma) > 0$, there exist mutually disjoint $M_i \in \mathcal{A}$ such that $\pi(\max M_i) < \pi(\gamma)$ and $M_i \cup \{\gamma\} \in \mathcal{A}$ for all $i \in \mathbb{N}$.

(iii) If $M \in \mathcal{A}$ and $\xi \in [0, \omega_1)$, then $M \cap ([0, \xi] \times [0, 1]) \in \mathcal{A}$.

(iv) If $M_i \in \mathcal{A}$, $i \in \mathbb{N}$, are pairwise disjoint and $0 < \xi < \omega_1$, there is $\gamma \in \Gamma$ such that $\pi(\gamma) > \xi$, $\pi(\max M_i) < \pi(\gamma)$, and $M_i \cup \{\gamma\} \in \mathcal{A}$ for each $i \in \mathbb{N}$.

Lemma 8.4.1.

We have

$$\mathcal{R} = \Big\{ N \subset \Gamma : \ \text{if } (\xi, t) \in N, \text{ then } N \cap ([0, \xi] \times [0, 1]) \in \mathcal{A}' \Big\}.$$

Moreover, every infinite $N \in \mathcal{R}$ is of the form $N = \{(\xi_1, t_1), (\xi_2, t_2), \ldots\}$, where $0 \leq \xi_1 < \xi_2 < \cdots$ and hence K is a Corson compact.

Proof. Take a set $N \subset \Gamma$ such that $N \cap ([0, \xi] \times [0, 1]) \in \mathcal{A}'$ whenever $(\xi, t) \in N$. If N is finite, then, clearly, $N \in \mathcal{A}'$ $(\subset \mathcal{R})$. Assume then that N is infinite. We may find a sequence $0 \leq \xi_1 < \xi_2 < \cdots < \omega_1$ such that $\#N \cap ([0, \xi_i] \times [0, 1]) = i$. Put $\xi = \lim_{i \to \infty} \xi_i$. If there were $(\eta, t) \in N$, with $\eta \geq \xi$, then we would get that the set $N \cap ([0, \eta] \times [0, 1])$ is infinite, a contradiction. Therefore $N \subset [0, \xi) \times [0, 1]$ and N must be of the form

$\{(\xi_1, t_1), (\xi_2, t_2), \ldots\}$ for some $t_1, t_2, \ldots \in [0, 1]$. It then also follows that $\chi_{N \cap ([0, \xi_i] \times [0, 1])} \to \chi_N$ as $i \to \infty$; thus $\chi_N \in K$ and $N \in \mathcal{R}$. We have proved the inclusion "⊃".

Now take $N \in \mathcal{R}$. If $\#N \le 1$, then we are done, so assume that $\#N \ge 2$. We may find a net $\{M_\tau\}$ in $\mathcal{A}'$ such that $\chi_{M_\tau} \to \chi_N$. Consider any distinct $\gamma_1, \gamma_2 \in N$, with $\pi(\gamma_1) \le \pi(\gamma_2)$. Then $M_\tau \supset \{\gamma_1, \gamma_2\}$ for large τ, and hence we must have that $\pi(\gamma_1) < \pi(\gamma_2)$. Now put $(\xi_1, t_1) = \min N$ and take any other $(\xi, t) \in N$. Then $\{(\xi_1, t_1), (\xi, t)\} \subset M_\tau$ for all large τ. Let $M'_\tau \in \mathcal{A}$ be such that $M'_\tau \supset M_\tau$ and $\pi(\min M'_\tau) = 0$; see the definition of $\mathcal{A}$. The property (i) guarantees that $M'_\tau \cap ([0, \xi] \times [0, 1])$ is a constant set and so

$$N \cap ([0, \xi] \times [0, 1]) = N \cap ([\xi_1, \xi] \times [0, 1]) = M_\tau \cap ([\xi_1, \xi] \times [0, 1]) \in \mathcal{A}'$$

for all large τ.

By putting together all the above steps, we can conclude that if N is an infinite element of $\mathcal{R}$, then it is of the form $N = \{\gamma_1, \gamma_2, \ldots\}$, where $0 \le \pi(\gamma_1) < \pi(\gamma_2) < \cdots$, which guarantees that K is Corson compact. □

Lemma 8.4.2.

The space K is Gul'ko (even Talagrand) compact but not Eberlein compact.

Proof. For $n, m \in \mathbb{N}$ let Γ_n^m be the set of all $\gamma \in \Gamma$ such that either there is $N \in \mathcal{A}$ with $\min N = (0, 1/n)$, $\max N = \gamma$ and $\#N = m$, or $\min N \ne (0, 1/n)$ for every $N \in \mathcal{A}$ satisfying $\max N = \gamma$. We shall verify the condition (vi) in Theorem 7.2.5. Take any $\gamma \in \Gamma$ and any infinite $M \in \mathcal{R}$. We find $M' \in \mathcal{R}$ and $n \in \mathbb{N}$ such that $\min M' = (0, 1/n)$. Then property (i) guarantees that $M' \cap \Gamma_n^m$ and hence $M \cap \Gamma_n^m$ is at most a singleton for every $m \in \mathbb{N}$. Further, if there is $N \in \mathcal{A}$ with $\max N = \gamma$ and $\min N = (0, 1/n)$, then $\gamma \in \Gamma_n^m$ for $m = \#N$. In the opposite case, $\gamma \in \Gamma_n^m$ for every $m \in \mathbb{N}$. Therefore, by Theorem 7.2.5, K is Gul'ko compact.

If we want to prove that K is even Talagrand compact, we put $\Gamma_\emptyset = \Gamma$, and $\Gamma_{\{m_1, \ldots, m_j\}} = \Gamma_1^{m_1} \cap \cdots \cap \Gamma_j^{m_j}$ for every finite sequence $m_1, \ldots, m_j$ of positive integers. This system of subsets of Γ satisfies condition (vi) from the end of Section 7.4, and hence K is Talagrand compact. In fact, if $M \in \mathcal{R}$ is infinite, we find M' and n as in the previous paragraph. Then for every $\sigma \in \mathbb{N}^{\mathbb{N}}$ we have

$$M \cap \Gamma_{\sigma|n} \subset M' \cap \Gamma_{\sigma|n} \subset M' \cap \Gamma_n^{\sigma(n)}$$

and the last set is a singleton. □

Assume now that K is Eberlein compact. Since $K \subset \{0, 1\}^\Gamma$ and every $\chi_M \in K$ can be written as $\lim_{n \to \infty} \chi_{M_n}$, where $\chi_{M_n} \in K$ and each M_n is finite, Theorem 4.3.2 guarantees the existence of sets $\Gamma_i \subset \Gamma$, $i = 1, 2, \ldots$, such that $\bigcup_{i=1}^\infty \Gamma_i = \Gamma$ and $M \cap \Gamma_i$ is finite for every $M \in \mathcal{R}$ and every $i \in \mathbb{N}$. However, according to the next lemma, this is not possible.

Lemma 8.4.3.

Let Γ_i, $i \in \mathbb{N}$, be subsets of a set Γ such that $\bigcup_{i=1}^{\infty} \Gamma_i = \Gamma$. Then there exist $i_0 \in \mathbb{N}$ and $M \in \mathcal{R}$ such that $M \cap \Gamma_{i_0}$ is an infinite set.

Proof. Replacing the sequence $\Gamma_1, \Gamma_2, \Gamma_3, \ldots$ by $\Gamma_1, \Gamma_2 \backslash \Gamma_1, \Gamma_3 \backslash (\Gamma_1 \cup \Gamma_2), \ldots$, we may and do assume that the sets Γ_i are pairwise disjoint. Let L be the set of all $i \in \mathbb{N}$ such that $\pi(\Gamma_i)$ is uncountable. Put $\xi = \sup \pi(\bigcup_{i \notin L} \Gamma_i)$; then $0 < \xi < \omega_1$. Assume first that L is finite. In this case, we may take $\chi_M \in K$ such that $M \cap ([\xi, \omega_1) \times [0, 1])$ is an infinite set; the existence of such M is guaranteed by the property (iv). Then

$$M \cap ([\xi, \omega_1) \times [0, 1]) = \bigcup_{i \in L} \Gamma_i \cap M \cap ([\xi, \omega_1) \times [0, 1]) \subset \bigcup_{i \in L} (\Gamma_i \cap M).$$

Therefore there is at least one $i_0 \in L$ such that $M \cap \Gamma_{i_0}$ is an infinite set.

Second, assume that L is infinite; write $L = \{m_1, m_2, \ldots\}$. We shall construct $M_1, M_2, \ldots$ in $\mathcal{A}$ such that $\max M_i \in \Gamma_{m_i}$ for every $i \in \mathbb{N}$ and $M_i \cap M_j = \emptyset$ if $i, j \in \mathbb{N}$ are distinct. Consider $n \in \mathbb{N}$ and assume we have already constructed $M_1, \ldots, M_{n-1}$ with the above properties. We pick some $\gamma \in \Gamma_{m_n}$, with $\pi(\gamma) > 0$. By property (ii) we find mutually disjoint sets $N_1, N_2, \ldots \in \mathcal{A}$ such that $\pi(\max N_i) < \pi(\gamma)$ and $N_i \cup \{\gamma\} \in \mathcal{A}$ for every $i \in \mathbb{N}$. Then necessarily $N_i \cap (M_1 \cup \cdots \cup M_{n-1}) = \emptyset$ for some $i \in \mathbb{N}$. (If $n = 1$, we take, say, $i = 1$.) Put then $M_n = N_i \cup \{\gamma\}$. This finishes the induction step.

Assume now that our lemma is false. Put $\eta = \sup \pi(\bigcup_{n=1}^{\infty} M_n)$; note that $\eta < \omega_1$. Then for every $n \in \mathbb{N}$ there is $M'_n \in \mathcal{A}$ such that $M'_n \supset M_n$, $M'_n \backslash M_n \subset \Gamma_{m_n} \cap ([\eta, \omega_1) \times [0, 1])$, and M'_n is maximal with respect to these properties. The sets M'_n are pairwise disjoint since so are M_n and Γ_{m_n}. By property (iv), there is $\gamma \in \Gamma$ such that $\pi(\gamma) > \xi$, $\pi(\max M'_n) < \pi(\gamma)$, and $M'_n \cup \{\gamma\} \in \mathcal{A}$ for all $n \in \mathbb{N}$. Then $\gamma \in \Gamma_{m_{n_0}}$ for some $n_0 \in \mathbb{N}$. This means M'_{n_0} has an extension $M'_{n_0} \cup \{\gamma\} \in \mathcal{A}$, which contradicts the maximality of M'_{n_0}. $\square$

Now we shall perform James's construction on Rezničenko's family $\mathcal{R}$. Denote

$$\Phi = \{x \in \mathbb{R}^{\Gamma} : \operatorname{supp} x \text{ is finite}\}.$$

For $x \in \Phi$ we put

$$\|x\| = \sup \left(\sum_{i=1}^{n} \left(\sum_{\gamma \in a_i} x(\gamma) \right)^2 \right)^{1/2},$$

where the supremum is taken over all finite families $a_1, \ldots, a_n$ of pairwise disjoint elements of $\mathcal{R}$. Thus, $(\Phi, \| \cdot \|)$ is a normed linear space. We shall denote by $J\mathcal{R}$ the completion of Φ. For $a \in \mathcal{R}$ we define $a^* : \Phi \to \mathbb{R}$ by $a^*(x) = \sum_{\gamma \in a} x(\gamma)$, $x \in \Phi$. Obviously, a^* is a continuous linear functional on

$(\Phi, \| \cdot \|)$. Let a^* also denote its extension to $J\mathcal{R}$. Clearly $\|a^*\| = 1$ whenever $\emptyset \neq a \in \mathcal{R}$. Let D denote the set of all sums $\sum_{i=1}^{n} \lambda_i a_i^*$, where $\sum_{i=1}^{n} \lambda_i^2 \leq 1$, and $a_1, \ldots, a_n$ is a family of pairwise disjoint elements of $\mathcal{R}$.

Lemma 8.4.4.

The weak* closure $\overline{D}^*$ of D contains all the extremal points of the unit ball $B_{J\mathcal{R}^*}$ of $J\mathcal{R}^*$.

Proof. Take any $\sum_{i=1}^{n} \lambda_i a_i^* \in D$. Then for $x \in \Phi$ we have

$$\left\langle \sum_{i=1}^{n} \lambda_i a_i^*, x \right\rangle = \sum_{i=1}^{n} \lambda_i \left(\sum_{\gamma \in a_i} x(\gamma) \right)$$

$$\leq \left(\sum_{i=1}^{n} \lambda_i^2 \right)^{1/2} \left(\sum_{i=1}^{n} \left(\sum_{\gamma \in a_i} x(\gamma) \right)^2 \right)^{1/2}$$

$$\leq \left(\sum_{i=1}^{n} \lambda_i^2 \right)^{1/2} \|x\|.$$

Hence

$$\left\| \sum_{i=1}^{n} \lambda_i a_i^* \right\| \leq \left(\sum_{i=1}^{n} \lambda_i^2 \right)^{1/2} \leq 1$$

and so $D \subset B_{J\mathcal{R}^*}$, which gives us that $\overline{\text{co } D}^* \subset B_{J\mathcal{R}^*}$.

Assume that there is $x^* \in B_{J\mathcal{R}^*} \setminus \overline{\text{co } D}^*$. By the separation theorem, we find $x \in J\mathcal{R}$, $\|x\| = 1$, and $\epsilon > 0$ such that $\langle x^*, x \rangle > \sup\langle D, x \rangle + 3\epsilon$. From the definition of $J\mathcal{R}$ we find $y \in \Phi$, $\|y\| = 1$ such that $\langle x^*, y \rangle > \sup\langle D, y \rangle + 2\epsilon$. We choose a pairwise disjoint family $a_1, \ldots, a_n$ of elements from $\mathcal{R}$ such that $\|y\| - \epsilon < \left(\sum_{i=1}^{n} \left(\sum_{\gamma \in a_i} y(\gamma) \right)^2 \right)^{1/2}$. Denote $\lambda_i = \sum_{\gamma \in a_i} y(\gamma)$, $i = 1, \ldots, n$, and $y^* = \sum_{i=1}^{n} \lambda_i a_i^*$. Thus

$$1 - \epsilon = \|y\| - \epsilon < \left(\sum_{i=1}^{n} \lambda_i^2 \right)^{1/2} = \langle y^*, y \rangle^{1/2}.$$

Also $\|y^*\| \leq \left(\sum_{i=1}^{n} \lambda_i^2 \right)^{1/2} \leq 1$ and hence $y^* \in D$. But then

$$(1 - \epsilon)^2 < \langle y^*, y \rangle \leq \sup\langle D, y \rangle < \langle x^*, y \rangle - 2\epsilon \leq 1 - 2\epsilon,$$

which is impossible.

In this way we have proved that $B_{J\mathcal{R}^*} = \overline{\text{co } D}^*$. Now $\overline{\text{co } D}^* = \overline{\text{co}\left(\overline{D}^*\right)}^*$ and so Milman's theorem [DS, Lemma V.8.5] finishes the proof. $\square$

Let E denote the set of all sequences $\{\chi_{a_n}\}$ where $\{a_1, a_2, \ldots\}$ are infinite families of mutually disjoint elements of $\mathcal{R}$. It is straightforward to verify that E, provided with the product topology, is a closed subspace of the compact $K^{\mathbb{N}}$. Further we define

$$\Omega = \left\{ \{\lambda_n\} \in B_{\ell_2} : |\lambda_1| \geq |\lambda_2| \geq \cdots \right\}$$

and endow it with the weak* topology. Finally, let Ω_0 be the subspace of Ω consisting of all sequences with finite support.

Lemma 8.4.5.

The mapping $T : \Omega \times E \to (J\mathcal{R}^*, w^*)$ defined by

$$T(\{\lambda_n\}, \{\chi_{a_n}\}) = \sum_{n=1}^{\infty} \lambda_n a_n^*, \qquad (\{\lambda_n\}, \{\chi_{a_n}\}) \in \Omega \times E,$$

is well defined and continuous and $\overline{D}^* = T(\Omega \times E) = \overline{D}$.

Proof. Take $(\{\lambda_n\}, \{\chi_{a_n}\}) \in \Omega \times E$. Since $\left\| \sum_{i=n}^{n+m} \lambda_i a_i^* \right\| \leq \left(\sum_{i=n}^{n+m} \lambda_i^2 \right)^{1/2}$, T is well defined.

As regards to the continuity of T, consider a net $\{z_\tau\} = \left\{ (\{\lambda_n^\tau\}, \{\chi_{a_n^\tau}\}) \right\}$ converging to $z = (\{\lambda_n, \}, \{\chi_{a_n}\})$. Then $\lambda_n^\tau \to \lambda_n$ and $\chi_{a_n^\tau} \to \chi_{a_n}$ for every $n \in \mathbb{N}$. Since the net $\{Tz_\tau\}$ is bounded and the functions $\chi_{\{\gamma\}}$, $\gamma \in \Gamma$, are linearly dense in $J\mathcal{R}$, it is enough to show that $\langle Tz_\tau, \chi_{\{\gamma\}} \rangle \to \langle Tz, \chi_{\{\gamma\}} \rangle$ for every $\gamma \in \Gamma$. So fix any $\gamma \in \Gamma$. Assume first that $\gamma \in a_m$ for some $m \in \mathbb{N}$. Then $\gamma \in a_m^\tau$ for τ large enough. For these τ we have

$$\langle Tz_\tau, \chi_{\{\gamma\}} \rangle = \left\langle \sum_{n=1}^{\infty} \lambda_n^\tau (a_n^\tau)^*, \chi_{\{\gamma\}} \right\rangle = \lambda_m^\tau \to \lambda_m = \langle Tz, \chi_{\{\gamma\}} \rangle.$$

Second, let $\gamma \in \Gamma \backslash \bigcup_{n=1}^{\infty} a_n$. Then clearly $\langle Tz, \chi_{\{\gamma\}} \rangle = 0$. We shall show that $\langle Tz_\tau, \chi_{\{\gamma\}} \rangle \to 0$. So let $\epsilon > 0$ be given. We find $m \in \mathbb{N}$ such that $|\lambda_m| < \epsilon$. Then we find τ_0 so large that $\gamma \notin a_1^\tau \cup \cdots \cup a_m^\tau$ and $|\lambda_m^\tau| < \epsilon$ whenever $\tau > \tau_0$. For such τ's we thus have

$$|\langle Tz_\tau, \chi_{\{\gamma\}} \rangle| = \left| \left\langle \sum_{n=m+1}^{\infty} \lambda_n^\tau (a_n^\tau)^*, \chi_{\{\gamma\}} \right\rangle \right|$$
$$\leq \sup \left\{ |\lambda_{m+1}^\tau|, |\lambda_{m+2}^\tau|, \ldots \right\} = |\lambda_{m+1}^\tau| \leq |\lambda_m^\tau| < \epsilon.$$

Let us prove the last statement of our lemma. Clearly, $T(\Omega \times E) \subset T(\Omega_0 \times E)$. Since T is continuous and $\Omega \times E$ is compact, $T(\Omega \times E)$ is weak* closed; thus $\overline{T(\Omega_0 \times E)}^* \subset T(\Omega \times E)$. Therefore

$$\overline{T(\Omega_0 \times E)}^* = T(\Omega \times E) = \overline{T(\Omega_0 \times E)},$$

and, as $D = T(\Omega_0 \times E)$, we are done. $\square$

Theorem 8.4.6.

The Banach space $J\mathcal{R}$ is weakly $\mathcal{K}$-analytic (hence Vašák) but is not isomorphic to a subspace of a WCG space or even to a subspace of an Asplund generated space. Moreover, $J\mathcal{R}$ does not contain ℓ_1 isomorphically.

Proof. From Lemma 8.4.2 we know that $K\ (= \{\chi_M : M \in \mathcal{R}\})$ is Talagrand compact. Hence, according to Theorem 7.1.10, so are the spaces $K^{\mathbb{N}}$, E, $\Omega \times E$, and finally $(\overline{D}, w^*)$; see the notation from Lemma 8.4.5. Lemma 8.4.4 guarantees that $J\mathcal{R}$ isometrically embeds into $C((\overline{D}, w^*))$. Thus, $J\mathcal{R}$ is a weakly $\mathcal{K}$-analytic space by Theorems 7.1.8 and 7.1.11.

Now, assume that $J\mathcal{R}$ is isomorphic to a subspace of an Asplund generated space. Then it is isomorphic to a subspace of a WCG space by Theorem 8.3.7. Consequently, by Theorems 1.2.5 and 1.2.4 and a theorem of Benyamini, M.E. Rudin, and Wage [BRW], $(B_{J\mathcal{R}^*}, w^*)$ is Eberlein compact and so is $(\overline{D}, w^*)$. Now, an inspection of the mapping T from Lemma 8.4.5 reveals that K occurs in $(\overline{D}, w^*)$ up to a homeomorphism. It follows, then, that K is Eberlein compact. However, this is not true according to Lemma 8.4.2.

It remains to prove the last statement of our theorem. Let $\{x_n\}$ be a sequence in $B_{J\mathcal{R}^*}$. According to Rosenthal's theorem [Di2, p. 201], we have to show that there exists a subsequence $\{x_{n_j}\}$ such that $\lim_{j\to\infty}\langle x^*, x_{n_j}\rangle$ exists for every $x^* \in J\mathcal{R}^*$. Since Φ is norm-dense in $J\mathcal{R}$, we may and do assume that every x_n has a finite support. By Rainwater's theorem [Ra] it is enough to consider x^*'s which are extremal points of $B_{J\mathcal{R}^*}$. Hence, by Lemma 8.4.4, we may restrict ourselves to $x^* \in \overline{D}\ (= \overline{D}^*)$ and, finally, to x^*'s of form a^*, where $a \in \mathcal{R}$.

We claim: *For every $\epsilon > 0$ and every infinite set $N \subset \mathbb{N}$ there exist a (possibly empty) finite family $a_1, \ldots, a_j$ of pairwise disjoint elements of $\mathcal{R}$ and an infinite set $N' \subset N$ such that*

$$\limsup_{n \in N'} |\langle a^*, x_n\rangle| \le \epsilon \quad \text{whenever} \quad a \in \mathcal{R} \quad \text{and} \quad a \cap (a_1 \cup \cdots \cup a_j) = \emptyset.$$

To prove this, fix ϵ and N. If $\limsup_{n\in N} |\langle a^*, x_n\rangle| \le \epsilon$ for every $a \in \mathcal{R}$, we are done. Otherwise, we choose $a_1 \in \mathcal{R}$ and an infinite set $N_1 \subset N$ satisfying $|\langle a_1^*, x_n\rangle| > \epsilon$ for all $n \in N_1$. Assume now that we have already found mutually disjoint sets $a_1, \ldots, a_j \in \mathcal{R}$ and infinite sets $N \supset N_1 \supset \cdots \supset N_j$ such that $|\langle a_i^*, x_n\rangle| > \epsilon$ for all $n \in N_i$ and all $i = 1, \ldots, j$. If $\limsup_{n\in N_j} |\langle a^*, x_n\rangle| \le \epsilon$ for all $a \in \mathcal{R}$ with $a \cap (a_1 \cup \cdots \cup a_j) = \emptyset$, then we are done. Otherwise we find $a_{j+1} \in \mathcal{R}$ and an infinite set $N_{j+1} \subset N_j$ such that $a_{j+1} \cap (a_1 \cup \cdots \cup a_j) = \emptyset$ and

$|\langle a_{j+1}^*, x_n \rangle| > \epsilon$ for all $n \in N_{j+1}$. This process must eventually stop. Indeed, take some $n \in N_j$ and find $\epsilon_i \in \{-1, 1\}$ such that $|\langle a_i^*, x_n \rangle| = \langle \epsilon_i a_i^*, x_n \rangle$, $i = 1, \ldots, j$. Then

$$j\epsilon < |\langle a_1^*, x_n \rangle| + \cdots + |\langle a_j^*, x_n \rangle| = \langle \epsilon_1 a_1^* + \cdots + \epsilon_j a_j^*, x_n \rangle$$
$$\leq \|\epsilon_1 a_1^* + \cdots + \epsilon_j a_j^*\| \leq \sqrt{j}$$

and therefore $j < \epsilon^{-2}$; we used here the fact that the sets $a_1, \ldots, a_j$ are pairwise disjoint and the estimate from the proof of Lemma 8.4.4.

Next, for every $m \in \mathrm{IN}$ we find, by the claim, an infinite set $N_m \subset \mathrm{IN}$ and a finite family $a_1^m, \ldots, a_{j_m}^m$ of mutually disjoint elements from $\mathcal{R}$ such that $\limsup_{n \in N_m} |\langle a^*, x_n \rangle| \leq 1/m$ whenever $a \in \mathcal{R}$ and $a \cap (a_1^m \cup \cdots \cup a_{j_m}^m) = \emptyset$. We may and do arrange the things in such a way that $N_1 \supset N_2 \supset \cdots$. Let $N_0 \subset \mathrm{IN}$ be an infinite set such that $N_0 \backslash N_m$ is finite for every $m \in \mathrm{IN}$.

By a diagonal argument, we may find an infinite set $N_0' \subset N_0$ such that $\lim_{n \in N_0'} \langle b^*, x_n \rangle$ exists whenever b belongs to the (countable) set

$$\left\{ \{\gamma\} : \gamma \in \bigcup_{m=1}^{\infty} \operatorname{supp} x_m \right\} \cup \{a_i^m : i = 1, \ldots, j_m, \ m \in \mathrm{IN}\};$$

then, clearly, $\lim_{n \in N_0'} \langle \{\gamma\}^*, x_n \rangle$ exists for every $\gamma \in \Gamma$. We also observe that $N_0' \cap N_m$ is an infinite set for every $m \in \mathrm{IN}$.

Now we are ready to prove that $\lim_{n \in N_0'} \langle a^*, x_n \rangle$ exists for every $a \in \mathcal{R}$. So fix an arbitrary $a \in \mathcal{R}$. Assume first that there exists $m \in \mathrm{IN}$ and $i \in \{1, \ldots, j_m\}$ such that $a \cap a_i^m$ is infinite. Using Lemma 8.4.1 and the definition of the family $\mathcal{A}'$, we find $c, d \in \mathcal{R}$ such that $\pi(\min c) = \pi(\min d) = 0$ and $c \supset a$, $d \supset a_i^m$. Then both $c \backslash a$ and $d \backslash a_i^m$ are finite sets and property (i) above Lemma 8.4.1 ensures that $c = d$. Hence

$$a \backslash a_i^m \subset c \backslash a_i^m = d \backslash a_i^m, \qquad a_i^m \backslash a \subset d \backslash a = c \backslash a,$$

so $(a \backslash a_i^m) \cup (a_i^m \backslash a)$ is a finite set. Now, observing that

$$a^* = (a_i^m)^* + \sum \left\{ \{\gamma\}^* : \gamma \in a \backslash a_i^m \right\} - \sum \left\{ \{\gamma\}^* : \gamma \in a_i^m \backslash a \right\},$$

we conclude that $\lim_{n \in N_0'} \langle a^*, x_n \rangle$ exists.

Second, assume that $a \cap a_i^m$ is finite for every $m \in \mathrm{IN}$ and every $i \in \{1, \ldots, j_m\}$. If a is finite, we are done. So let us assume that a is infinite. Take any $\epsilon > 0$. We may find $m \in \mathrm{IN}$ such that $2/m < \epsilon$. Write $a = \{\gamma_1, \gamma_2, \ldots\}$, where $\pi(\gamma_1) < \pi(\gamma_2) < \cdots$; see Lemma 8.4.1. We find $i \in \mathrm{IN}$ such that $c := \{\gamma_i, \gamma_{i+1}, \ldots\}$ is disjoint from $a_1^m \cup \cdots \cup a_{j_m}^m$. However, since $c \in \mathcal{R}$, we have by the claim

$$0 \leq \limsup_{n \in N_0'}\langle a^*, x_n \rangle - \liminf_{n \in N_0'}\langle a^*, x_n \rangle$$

$$= \limsup_{n \in N_0'}\langle c^*, x_n \rangle - \liminf_{n \in N_0'}\langle c^*, x_n \rangle$$

$$\leq \limsup_{n \in N_m}\langle c^*, x_n \rangle - \liminf_{n \in N_m}\langle c^*, x_n \rangle \leq \frac{2}{m} < \epsilon.$$

Here $\epsilon > 0$ was arbitrary. Therefore $\lim_{n \in N_0'}\langle a^*, x_n \rangle$ exists.   □

8.5.   NOTES AND REMARKS

Theorem 8.1.2 is due to Jayne and Rogers. For an account on selection theorems for upper semicontinuous multivalued mappings see [JR, HJT, GMS] and the references therein. Perhaps the first result of this type for monotone operators appeared in [F1]. Our proof of Theorem 8.1.2, which is based on very old ideas, is due to Stegall [St11]. Lemma 8.1.3, whose proof is based on ideas of R.C. James and J.D. Pryce, is due to Simons [Si]; see also [R].

Proposition 8.2.1 is from [OV]. Its proof is based on an argument from [FG]; see also [St9, St10]. Theorem 8.2.2 is from [FG]. In [St13], it is shown how to avoid Lemma 8.1.3 from its proof. This result is a culmination of a sequence of papers [L1, L2, AL, T, JZ1, F2, F3] started by Lindenstrauss. As a by-product, it is shown in [FG] that *Banach spaces which are M-ideals in their biduals are WCG and contain a complemented copy of c_0.*

Proposition 8.3.1 is from Valdivia [Va1]. He assumed that V is WLD instead of the Corsonness of (B_{V^*}, w^*). However, it should be noted that these two conditions are equivalent [OSV]. For Proposition 8.3.2, when $Y = V$, see Orihuela and Valdivia [OV]. Its proof is based on ideas of John and Zizler [JZ2] and on [F3]. It should be noted that [F3] is the first paper where the powerful tool of Baire class 1 selectors for usco mappings in the context of P.R.I. appeared. In Theorem 8.3.3, (i)⇔(ii)⇔(iii)⇔(iv) is from [F3] and (i)⇔(v) is from [Va1]. The equivalence (i)⇔(iv) is contained in (i)⇔(v). However, in order to show that (iv) implies (v), we need all the machinery of the P.R.I.; see Section 7. An account of predecessors of this theorem can be found in [Di1, Chapter 4., §§ 7–9]. From Theorems 1.1.3 and 8.3.3 we get that scattered compacta which are also Gul'ko, or only Corson, are Eberlein [F3, Al].

There are two papers [St10, OSV], which appeared approximately at the same time, where Theorems 8.3.4 and 8.3.5 can be found. However, we heard that Theorem 8.3.5 had already been proved, a few years prior to these papers, by E. A. Rezničenko. For yet another proof of Theorem 8.3.4, without use of any interpolation, see [FW]. Once we have at hand the result of S. P. Gul'ko, E. Michael, and M. E. Rudin that *the continuous image of a Corson compact is Corson compact* [Ar, Corollary IV.3.15], we may simplify our proof of

Theorem 8.3.4. In fact, we then know that (B_{Y^*}, w^*) is Corson compact and we may use Proposition 8.3.2 in the special form when $Y = V$. We then only need Proposition 6.1.10 for the situation when $Y = V$, and this case is easier to handle.

Theorem 8.3.6 is due to Stegall [St10]. It should be noted that the space $C(K)$ constructed in Section 4.3 is, by Theorems 4.3.1 and 8.3.7, an example of *a weakly $\mathcal{K}$-analytic space (hence with strictly convex dual norm [Me]) which is not a subspace of an Asplund generated space*. Further, if we put $V = C(K) \times C_0(T)$, where $C_0(T)$ is Haydon's example mentioned in Section 5.4, then (V^*, w^*) is fragmentable by Theorems 4.3.1, 7.2.8, 5.2.3, and 5.2.4(v). However, V is not Gateaux smoothable since $C_0(T)$ is not and V is not a subspace of an Asplund generated space since $C(K)$ is not; see Theorem 1.3.6(ii).

For further gymnastics with applications of the P.R.I. we refer to [DG]. Let us mention at least one result from here. A compact K is called a *Valdivia* compact if K can be found, up to a homeomorphism, in $\mathbb{R}^\Gamma$ for some Γ and $K \cap \Sigma(\Gamma)$ is dense in K [Va2, DG]. A canonical example is the long interval $[0, \omega_1]$, where ω_1 is the first uncountable ordinal. (A canonical counterexample is the interval $[0, \alpha]$, where card $\alpha > \aleph_1$.) It is easy to show that this space is Valdivia compact. It is also a Radon–Nikodým compact since $C([0, \omega_1])$ is, by Theorem 1.1.3, an Asplund space. Yet $[0, \omega_1]$ is not Eberlein compact; see the text below Proposition 1.5.2. Hence the analogue of Theorem 8.3.5 does not hold. It is shown in [DG] that *Corson compacta are exactly those Valdivia compacta which do not contain, homeomorphically,* $[0, \omega_1]$.

We cannot help mentioning some renorming consequences of the results of this chapter. *The dual of every Asplund space admits an equivalent (not necessarily dual) LUR norm* [FG; DGZ2, Corollary VII.1.12]. *A WCG Asplund space admits an equivalent norm such that it as well as its dual norm is LUR* [F3; DGZ2, Theorem VII.1.14]. *A Banach space admits an equivalent LUR norm if its dual unit ball with weak* topology is Corson compact.* The proofs are based on P.R.I., on Troyanski's deep renorming method [DGZ2, Section VII.1], eventually on Zizler's variant of it [Z1], as well as on Asplund's averaging technique [DGZ2, Theorem II.4.1].

In Section 8.4, we followed the paper of Argyros [Arg]. If we replace $\mathcal{R}$ by the dyadic tree $\mathcal{D}$, then we get James's original construction: *$J\mathcal{D}$ is a separable non-Asplund space not containing an isomorphic copy of ℓ_1* [J]. When considering a Todorčević tree $\mathcal{T}$ [To] instead of $\mathcal{R}$, we get a *space $J\mathcal{T}$ which is not weakly $\mathcal{K}$-analytic and does not contain an isomorphic copy of ℓ_1 but whose dual unit ball is Corson compact (i.e., $J\mathcal{T}$ is WLD)* [Arg]; this gives a counterexample related to (vi)$\Rightarrow$(iii) in Theorem 8.3.3. A combination of the techniques of Section 8.4 with the "Eberleinization" presented in Section 1.6 yields a *non-WCG space which is a subspace of a WCG space and which does not contain ℓ_1 isomorphically* [Arg]; this is a sort of strengthening of what we constructed in Section 1.6. Thus we have a "counterexample" to (ii)$\Rightarrow$(i) in Theorem 8.3.3. Probably there are no counterexamples to the implication (iv)⇒(iii) in the framework of spaces not containing ℓ_1. A natural question is, what would happen if we replaced $\mathcal{R}$ by the family $\mathcal{A}$ from the construction of Talagrand's compact in Section 4.3? Unfortunately, the family $\mathcal{A}$ is adequate and this easily implies that the corresponding James space contains ℓ_1.

Main Open Questions and Problems

To find any equivalent characterization for the weak Asplundness and for being a Gâteaux differentiability space.

Is every Gâteaux differentiability space weak Asplund?

To answer the above at least for spaces of the type $C(K)$.

If V is a weak Asplund space, is $V \times \mathbb{R}$ also weak Asplund?

If V is a weak Asplund (a Gâteaux differentiability) space, is every uncomplemented subspace of V also one?

Can we construct a space in $\mathcal{S} \backslash \mathcal{F}$ without any additional set-theoretical assumption? (See Definitions 3.13 and 5.0.1.)

If a compact K belongs to $\mathcal{S}$, is then $C(K)$ in $\widetilde{\mathcal{S}}$? (See Definition 3.2.1.)

Is $\widetilde{\mathcal{S}} = \widetilde{\mathcal{F}}$? (See Definition 5.2.1.)

If V is a weak Asplund (a Gâteaux differentiability) space, is $C\big((B_{V^*}, w^*)\big)$ also one?

References

[Al] K. Alster, *Some remarks on Eberlein compacts,* Fundamenta Math. **104**(1979), 43–46.

[AP] K. Alster and R. Pol, *A function space C(K) which is weakly Lindelöf but not weakly compactly generated,* Studia Math. **64**(1979), 279–285.

[AL] D. Amir and J. Lindenstrauss, *The structure of weakly compact sets in Banach spaces,* Ann. Math. **88**(1968), 35–46.

[Ar] A. V. Archangel'skij, *Topological Spaces of Functions,* Moscow State Univ. 1989, in Russian, Dordrecht: Kluver Academic, 1992, in English.

[Arg] S. Argyros, *Weakly Lindelöf determined Banach spaces not containing ℓ_1,* Preprint, University of Athens.

[AM] S. Argyros and S. Mercourakis, *On weakly Lindelöf Banach spaces,* Rocky Mountain J. Math. **23**(1993), 395–446.

[AMN] S. Argyros, S. Mercourakis, and S. Negrepontis, *Functional-analytic properties of Corson compact spaces,* Studia Math. **89**(1988), 197–229.

[As] E. Asplund, *Fréchet differentiability of convex functions,* Acta Math. **121**(1968), 31–47.

[BRW] Y. Benyamini, M.E. Rudin, and M. Wage, *Continuous images of weakly compact subsets of Banach spaces,* Pacific J. Math. **70**(1977), 309–324.

[B] J. M. Borwein, *Weak local supportability and application to approximations,* Pacific J. Math. **82**(1979), 323–338.

[BP] J. M. Borwein and D. Preiss, *A smooth variational principle with applications to subdifferentiability and differentiability of convex functions,* Trans. Amer. Math. Soc. **303**(1987), 517–527.

[BR] J. Bourgain and H. P. Rosenthal, *Martingales valued in certain subspaces of L^1,* Israel J. Math. **37**(1980), 54–75.

[Bo] R. D. Bourgin, *Geometric Aspects of Convex Sets with the Radon-Nikodým Property,* Lecture Notes in Math. No. **993**, Berlin: Springer-Verlag, 1983.

[Č] E. Čech, *Point Sets,* Prague: Academia, 1968.

[Ch1] G. Choquet, *Theory of capacities,* Ann. Inst. Fourier **5**(1953), 131–297.

[Ch2] G. Choquet, *Ensembles K-analytiques and K-sousliniens, Cas général et cas métrique,* Ann. Inst. Fourier **9**(1959), 75–89.

[Chr] J. P. R.Christensen, *Theorems of Namioka and R.E. Johnson type for upper semicontinuous and compact valued set valued mappings,* Proc. Amer. Math. Soc. **86**(1982), 649–655.

[ChK] J. P. R. Christensen and P. S. Kenderov, *Dense strong continuity of mappings and the Radon-Nikodým property,* Math. Scandinavica **54**(1984), 70–78.

[Či] P. Čížek, *On some generalizations of Eberlein compacta,* Diploma Thesis, Charles University, Prague, 1996.

[ČK] M. M. Čoban and P. S. Kenderov, *Generic Gâteaux differentiability of convex functionals in C(T) and the topological properties of T,* Proc. of 15th Spring Conf. of Union of Bulgarian Math., Sljančev Brjag 1986, 141–149.

[DFJP] W. J. Davis, T. Figiel, W. B. Johnson, and A. Pełczyński, *Factoring weakly compact operators,* J. Funct. Analysis, **17**(1974), 311–327.

[De] G. Debs, *Espaces K-analytiques et espaces de Baire de fonctions continues,* Mathematika **32**(1985), 218–228.

[DG] R. Deville and G. Godefroy, *Some applications of projective resolutions of identity,* Bull. London Math. Soc. **67**(1993), 183–199.

[DGZ1] R. Deville, G. Godefroy, and V. E. Zizler, *Un principle variationel utilisant des fonctions bosses,* C.R. Acad. Sci. Paris, Série I, **312**(1991), 281–286.

[DGZ2] R. Deville, G. Godefroy, and V. E. Zizler, *Smoothness and Renormings in Banach Spaces,* Pitman Monographs No. **64**, London: Longman, 1993.

[Di1] J. Diestel, *Geometry of Banach Spaces — Selected Topics,* Lecture Notes in Math. No. **485**, Berlin: Springer-Verlag, 1975.

[Di2] J. Diestel, *Sequences and Series in Banach Spaces,* GTM 92, Berlin: Springer-Verlag, 1984.

[DU] J. Diestel and J. J. Uhl, Jr., *Vector Measures,* Amer. Math. Soc. Surveys No. **15**, Providence, Rhode Island, 1977.

[DN] D. van Dulst and I. Namioka, *A note on trees in conjugate Banach spaces,* Indag. Math. **46**(1984), 7–10.

[DS] N. Dunford and J. T. Schwartz, *Linear Operators, Part I,* New York: Interscience, 1958.

[EL] I. Ekeland and G. Lebourg, *Generic Fréchet differentiability and perturbed optimization problems in Banach spaces,* Trans. Amer. Math. Soc. **224**(1976), 193–216.

[En] R. Engelking, *General Topology,* Warszawa: PWN, 1985.

[F1] M. Fabian, *On strong continuity of monotone mappings in reflexive Banach spaces,* Acta Polytech. **IV, 2**(1976), 85–98.

[F2] M. Fabian, *On projectional resolution of identity on the duals of certain Banach spaces,* Bull. Australian Math. Soc. **35**(1987), 363–372.

[F3] M. Fabian, *Each weakly countably determined Asplund space admits a Fréchet differentiable norm,* Bull. Australian Math. Soc. **36**(1987), 367–374.

[F4] M. Fabian, *On dual locally uniformly rotund norm on a dual Vašák space,* Studia Math. **101** (1991), 69–81.

[FG] M. Fabian and G. Godefroy, *The dual of every Asplund space admits a projectional resolution of the identity,* Studia Math. **91**(1988), 141–151.

[FP] M. Fabian and D. Preiss, *On intermediate differentiability of Lipschitz functions on certain Banach spaces*, Proc. Amer. Math. Soc. **113**(1991), 733–740.

[FT] M. Fabian and S. L. Troyanski, *A Banach space admits a locally uniformly rotund norm if its dual is a Vašák space*, Israel J. Math. **69**(1990), 214–224.

[FW] M. Fabian and J. H. M. Whitfield, *On equivalent characterization of weakly compactly generated Banach spaces*, Rocky Mountains J. Math. **24**(1994), 1363–1378.

[FZ] M. Fabian and N. V. Živkov, *A characterization of Asplund spaces with help of local ε-support of Ekeland and Lebourg*, C. R. Acad. Bulgare Sci. **38**(1985), 671–674.

[Fa] V. Farmaki, *The structure of Eberlein, uniformly Eberlein and Talagrand compact spaces in* $\Sigma(\mathbb{R}^\Gamma)$, Fundamenta Math. **128**(1987), 15–28.

[Fi] S. P. Fitzpatrick, *Separably related sets and the Radon-Nikodým property*, Illinois J. Math. **29**(1985), 229–247.

[Fo] G. Fodor, *On stationary sets and regressive functions*, Acta Sci. Math. Szeged **27**(1966), 105–110.

[Fos] M. Fosgerau, *A Banach space with Lipschitz Gâteaux smooth bump has weak* fragmentable dual*, Part of author's Thesis. University College, London.

[Fr] Z. Frolík, *A survey of separable descriptive theory of sets and spaces*, Czechoslovak J. Math. **20**(1970), 406–467.

[GMS] N. Ghoussoub, B. Maurey, and W. Schachermayer, *Slicings, selections and their applications*, Canadian J. Math. **44**(1992), 483–504.

[Gi] J. R. Giles, *Convex Analysis with Application in Differentiation of Convex Functions*, Research Notes in Math. No. **58**, Boston: Pitman, 1982.

[Go1] G. Godefroy, *Compacts de Rosenthal*, Pacific J. Math. **91**(1980), 293–306.

[Go2] G. Godefroy, *Boundaries of a convex set and interpolation sets*, Math. Annalen **227**(1987), 173–184.

[Gow1] W. T. Gowers, *A solution to Banach's hyperplane problem*, Bull. London Math. Soc. **26**(1994), 523–530.

[Gow2] W. T. Gowers, *Recent Results in the Theory of Infinite-Dimensional Banach Spaces*, Proc. IMC, Zürich: Birkhäuser, 1994, pp 933–942.

[GM] W. T. Gowers and B. Maurey, *The unconditional basis sequence problem*, J. Amer. Math. Soc. **6**(1993), 851–874.

[G] A. Grothendieck, *Produits tensoriels et espaces nucleaires*, Memoirs Amer. Math. Soc. **16**, Providence, Rhode Island, 1955.

[Gr] G. Gruenhage, *A note on Gul'ko compact spaces*, Proc. Amer. Math. Soc. **100** (1987), 371–376.

[Gu] S. P. Gul'ko, *On the structure of spaces of continuous functions and their complete paracompactness*, Russian Math. Surveys **34**(1979), 36–44; Uspekchi Mat. Nauk **34**(1979), 33–40.

[HS] J. Hagler and F. E. Sullivan, *Smoothness and weak* sequential compactness*, Proc. American. Math. Soc. **78**(1980), 497–503.

[HHZ] P. Habala, P. Hájek, and V. Zizler, *Introduction to Banach Spaces*, Prague: Charles University Press, 1996.

[HJT] R. W. Hansell, J. E. Jayne, and M. Talagrand, *First class selectors for weakly upper semi-continuous multi-valued maps in Banach spaces*, J. Reine Ang. Mathematik **361**(1985), 201–220.

[Ha1] R. Haydon, *A counterexample to several questions about scattered compact spaces*, Bull. London Math. Soc. **22**(1990), 261–268.

[Ha2] R. Haydon, *Normes infiniment différentiables sur certains espaces de Banach*, C. R. Acad. Sci. Paris **315**(1992), 1175–1178.

[Ha3] R. Haydon, *The three-space problem for strictly convex renormings and the quotient problem for Fréchet-smooth renormings*, Séminaire Initiation à l'Analyse, Université Paris VI (1992/1993).

[Ha4] R. Haydon, *Baire trees, bad norms and the Namioka property*, Mathematika **42**(1995), 30–42.

[Ha5] R. Haydon, *Smooth functions and partitions of unity on certain Banach spaces*, to appear in Quart. J. Math. (Oxford).

[Ha6] R. Haydon, *Trees in renorming theory*, Preprint, Oxford University, England.

[HM] S. Heinrich and P. Mankiewicz, *Applications of ultra-powers to the uniform and Lipschitz classification of Banach spaces*, Studia Math. **73**(1982), 225–251.

[H] M. Heisler, *Singlevaluedness of monotone operators on subspaces of GSG spaces*, to appear in Comment. Math. Univ. Carolinae, **37**(1996), 255–261.

[HŠZ] P. Holický, M Šmídek, L. Zajíček, *Convex functions with nonmeasurable set of Gâteaux differentiability points*, Preprint, Charles's University, Prague.

[J] R. C. James, *A separable somewhat reflexive Banach space with non-separable dual*, Bull. American Math. Soc. **80**(1974), 738–743.

[JR] J. E. Jayne and C. A. Rogers, *Borel selectors for upper semicontinuous maps*, Acta Math. **155**(1985), 41–79.

[JZ1] K. John and V. Zizler, *Duals of Banach space which admit nontrivial smooth functions*, Bull. Australian Math. Soc. **11**(1974), 161–166.

[JZ2] K. John and V. Zizler, *Smoothness and its equivalents in weakly compactly generated Banach spaces*, J. Funct. Anal. **15**(1974), 1–11.

[JZ3] K. John and V. Zizler, *Markuševič bases in some dual spaces*, Proc. Amer. Math. Soc. **50**(1975), 293–296.

[JL1] W. B. Johnson and J. Lindenstrauss, *Some remarks on weakly compactly generated Banach spaces*, Israel J. Math. **17**(1974), 219–230.

[JL2] W. B. Johnson and J. Lindenstrauss, *Correction to "Some remarks on weakly compactly generated Banach spaces,"* Israel J. Math. **17**(1974), 219–230, Israel J. Math. **32**(1979), 382–383.

[Jo] L. Jokl, *Upper semicontinuous compact valued correspondences and Asplund spaces*, Preprint, Palacký University, Olomouc.

[KN] J. L. Kelley and I. Namioka, *Topological Linear Spaces*, GTM Princeton Univ. Press., 1963.

[K1] P. S. Kenderov, *The set-valued monotone mappings are almost everywhere single-valued*, Compt. Rendus Acad. Bulgare Sci. **27**(1974), 1173–1176.

[K2] P. S. Kenderov, *Monotone operators in Asplund spaces*, C.R. Acad. Bulgare Sci. **30**(1977), 963–964.

[K3] P. S. Kenderov, *Most optimization problems have unique solution*, International Series in Numerical Mathematics **72** Basel: Birkhauser, 1984, pp. 203–215.

[K4] P. S. Kenderov, *C(T) is wak Asplund for every Gul'ko compact T*, C. R. Acad. Bulgare Sci. **40**(1987), 17–20.

[K5] P. S. Kenderov, *A Talagrand compact space which is not a Radon-Nikodým space (An example of Rezničenko)*, Unpublished manuscript. Math. Institute, BAN, Sofia.

[KM] P. S. Kenderov and W. B Moors, *Game characterization of fragmentability of topological spaces*, Proc. of 25th Spring Conf. of the Union of Bulgarian Math., 1996, Kazanlak.

[KO] P. S. Kenderov and J. Orihuela, *A generic factorization theorem*, Mathematika, **42**(1995), 56–66.

[Ku] K. Kuratowski, *Topology I, II*, New York: Academic, 1966.

[KT] D. N. Kutzarova and S. L. Troyanski, *Reflexive Banach spaces without equivalent norms which are uniformly convex or uniformly differentiable in every direction*, Studia Math. **72**(1982), 92–95.

[LP] D. G. Larman and R. R. Phelps, *Gâteaux differentiability of convex functions on Banach spaces*, J. London Math. Soc. **20**(1979), 115–127.

[Le] A. G. Leiderman, *On everywhere dense metrizable subspaces of Corson compacta*, Matem. Zametki, **38**(1985), 440–449, in Russian.

[LS] A. G. Leiderman and G. A. Sokolov, *Adequate families of sets and Corson compacta*, Comment. Math. Univ. Carolinae **25**(1984), 233–246.

[L1] J. Lindenstrauss, *On reflexive spaces having the metric approximation property*, Israel J. Math. **3**(1965), 199–204.

[L2] J. Lindenstrauss, *On nonseparable reflexive Banach spaces*, Bull. Amer. Math. Soc. **72**(1966), 967–970.

[LT] J. Lindenstrauss and L. Tzafriri, *Classical Banach Spaces I*, Berlin: Springer-Verlag, 1977.

[Me] S. Mercourakis, *On weakly countably determined Banach spaces*, Trans. Amer. Math. Soc. **300**(1987), 307–327.

[MN] S. Mercourakis and S. Negrepontis, *"Banach spaces and topology II,"* Recent Progress in General Topology, M. Hušek and J. van Mill (eds.), 1992, pp 493–536.

[N1] I. Namioka, *Radon-Nikodým compact spaces and fragmentability*, Mathematika **34**(1987), 258–281.

[N2] I. Namioka, *Eberlein and Radon-Nikodým compact spaces*, Lecture Notes at University College London, Autumn 1985.

[NP] I. Namioka and R. R. Phelps, *Banach spaces which are Asplund spaces*, Duke Math. J. **42**(1975), 735–750.

[NW] I. Namioka and R. F. Wheeler, *Gul'ko's proof of the Amir-Lindenstrauss theorem*, Proc. M.M. Day Conference, Univeristy of Illinois, 1983.

[Ne] S. Negrepontis, *"Banach spaces and topology,"* Handbook of Set-Theoretic Topology, K. Kunen and J. E. Vaughan (eds.), Amsterdam: North-Holland, 1984, pp. 1041–1138.

[Ni] O. Nikodým, *Sur une proprieté de l'operation A*, Fundamenta Math. **7**(1925), 149–154.

[Ox] J. C. Oxtoby, *Measure and category*, Berlin: Springer-Verlag, 1980.

[O] J. Orihuela, *On Weakly Lindelöf Banach Spaces*, Progress in Funct. Anal., K. D. Bierstedt, J. Bonet, J. Horváth and M. Maestre (eds.), Elsevier, 1992, pp. 279–291.

[OSV] J. Orihuela, W. Schachermayer, and M. Valdivia, *Every Radon-Nikodým Corson compact is Eberlein compact*, Studia Math. **98**(1991), 157–174.

[OV] J. Orihuela and M. Valdivia, *Projective generators and resolutions of identity in Banach spaces*, Rev. Mat. Univ. Complutense, Madrid 2, Suppl. Issue (1990), 179–199.

[Pl] A. N. Pličko, *On projectional resolutions, Markuševič bases and equivalent norms*, Matem. Zametki **34**(1983), 719–726, in Russian.

[Ph] R. R. Phelps, *Convex Functions, Monotone Operators, and Differentiability*, Lecture Notes in Math. No. **1364**, 2nd Edition, Berlin: Springer-Verlag, 1993.

[Po] R. Pol, *Note on a theorem of S. Mercourakis about weakly K-analytic Banach spaces*, Comment. Math. Univ Carolinae **29**(1988), 723–730.

[Pr] D. Preiss, *Fréchet derivatives of Lipshitz functions*, J. Funct. Analysis **91**(1990), 312–345.

[PPN] D. Preiss, R. R. Phelps, and I. Namioka, *Smooth Banach spaces, weak Asplund spaces and monotone or usco mappings*, Israel J. Math. **72**(1990), 257-279.

[Ra] J. Rainwater, *Weak convergence of bounded sequences*, Proc. Amer. Math. Soc. **14**(1960), 999.

[Re1] O. I. Reinov, *Operators of type RN in Banach spaces*, Dokl. Acad. Nauk SSSR **220**(1975), 528–531.

[Re2] O. I. Reinov, *The Radon-Nikodým property and integral representations of linear operators*, Funkts. analiz i ego prilozhen. **9**(1975), 87–88.

[Re3] O. I. Reinov, *Some classes of sets in Banach spaces and a topological characterization of operators of type RN*, Zapiski nauchn. sem. LOMI **73**(1977), 224–228.

[Re4] O. I. Reinov, *Operators of type RN in Banach spaces*, Sib. Mat. Ž. **19**(1978), 857–865.

[Re5] O. I. Reinov, *"On two questions in the theory of linear operators,"* Primenenie funktsional'nogo analiza *v teorii priblizhenij*, KGU, Kalinin, 1979, pp. 102–114.

[Re6] O. I. Reinov, *On a class of universally measurable mappings*, Matemat. zametki **26**(1979), 949–954.

[Re7] O. I. Reinov, *On integral representations of linear operators acting from the space L^1*, Matemat. zametki **27**(1980), 283–290.

[Re8] O. I. Reinov, *On a class of Hausdorff compacts and GSG Banach spaces*, Studia Math. **71** (1981), 113–126.

[Ri1] N. K. Ribarska, *Internal characterization of fragmentable spaces*, Mathematika **34** (1987), 243–257.

[Ri2] N. K. Ribarska, *The dual of a Gâteaux smooth Banach space is weak* fragmentable*, Proc. Amer. Math. Soc. **114**(1992), 1003–1008.

[RR] A. P Robertson and W. Robertson, *Topological Vector Spaces*, Cambridge, MA: Cambridge Univ. Press, 1964.

[R] R. Rodé, *Superkonvexität und schwache Kompaktheit*, Arch. Math. **36**(1981), 62–72.

[Ro] H. P. Rosenthal, *The heredity problem for weakly compactly generated Banach spaces*, Compositio Math. **28**(1974), 83–111.

[Ru1] W. Rudin, *Functional Analysis*, New York: McGraw-Hill, 1973.

[Ru2] W. Rudin, *Real and Complex Analysis*, New York: McGraw-Hill, 1974.

[Si] S. Simons, *A convergence theorem with boundary*, Pacific J. Math. **40**(1972), 703–708.

[SY] B. Sims and D. Yost, *Linear Hahn-Banach extension operators*, Proc. Edinburgh Math. Soc. **32**(1989), 53–57.

[So] G. A. Sokolov, *On some classes of compact spaces lying in Σ-products*, Comment. Math. Univ. Carolinae **25**(1984), 219–231.

[St1] Ch. Stegall, *The Radon-Nikodým property in conjugate Banach spaces*, Trans. Amer. Math. Soc. **206**(1975), 213–223.

[St2] Ch. Stegall, *The duality between Asplund spaces and spaces with the Radon-Nikodým property*, Israel J. Math. **29**(1978), 408–412.

[St3] Ch. Stegall, *The Radon–Nikodým property in conjugate Banach spaces II*, Trans. Amer. Math. Soc. **264**(1981), 507–519.

[St4] Ch. Stegall, *A class of topological spaces and differentiability*, Vorlesungen aus dem Fachbereich Mathematik der Universität Essen **10**(1983), 63–77.

[St5] Ch. Stegall, *"Gâteaux differentiation of functions on a certain class of Banach spaces,"* Funct. Analysis: Surveys and Recent Results III North Holland, 1984, pp. 35–46.

[St6] Ch. Stegall, *"More Gâteaux differentiability spaces,"* Banach Spaces, Proceedings Missouri 1984, N. Kalton and E. Saab (eds.), Lecture Notes No. 1166, Berlin: Springer-Verlag, 1985, pp. 186–199.

[St7] Ch. Stegall, *Topologiacal spaces with dense subspaces that are homeomorphic to complete metric spaces and the classification of $C(K)$ Banach spaces*, Mathematika **34**(1987), 101–107.

[St8] Ch. Stegall, *A proof of the theorem of Amir and Lindenstrauss*, Israel J. Math. **68**(1989), 185–192.

[St9] Ch. Stegall, *More facts about conjugate Banach spaces with the Radon-Nikodým property*, Acta Univ. Carolinae–Math. et Phys. **31**(1990), 107–117.

[St10] Ch. Stegall, *More facts about conjugate Banach spaces with the Radon-Nikodým property II*, Acta Univ. Carolinae–Math. Phys. **32**(1991), 47–54.

[St11] Ch. Stegall, *Functions of the first Baire class with values in Banach spaces*, Proc. Amer. Math. Soc. **111**(1991), 981–991.

[St12] Ch. Stegall, *The topology of certain spaces of measures*, Topology and Its Applications **41**(1991), 73–112.

[St13] Ch. Stegall, *"Spaces of Lipschitz functions on Banach spaces,"* Functional analysis (Essen 1991), 265–278, Lect. Notes in Pure and Appl. Math., 150, New York: Dekker, 1996.

[St14] Ch. Stegall, *A few remarks about our paper "The topology of certain spaces of measures,"* Preprint, J. Kepler University, Linz, 1995.

[T] D. G. Tacon, *The conjugate of a smooth Banach space*, Bull. Australian Math. Soc. **2**(1970), 415–425.

[Ta1] M. Talagrand, *Sur une conjecture de H.H. Corson*, Bull. Soc. Math. Série **99** (1975), 211–212.

[Ta2] M. Talagrand, *Espaces de Banach faiblement K-analytiques*, Annals Math. **110** (1979), 407–438.

[Ta3] M. Talagrand, *Deux examples de fonctions convexes*, C. R. Acad. Sci. Paris **288**(1979), 461–464.

[Ta4] M. Talagrand, *A new countably determined Banach space*, Israel J. Math. **47**(1984), 75–80.

[Ta5] M. Talagrand, *Renormages des quelques C(K),* Israel J. Math. **54**(1986), 327–334.

[Tay] E. A. Taylor, *Functional Analysis,* New York: Wiley, 1967.

[To] S. Todorčević, *"Trees and linearly ordered sets,"* Handbook of Set-Theoretic Topology, K. Kunen and J. E. Vaughan (eds.), Amsterdam: North-Holland, 1984, pp. 235–293.

[Tr1] S. L. Troyanski, *On locally uniformly convex and differentiable norms in certain non-separable Banach spaces,* Studia Math. **37**(1971), 173–180.

[Tr2] S. L. Troyanski, *On equivalent norms and minimal systems in nonseparable Banach spaces,* Studia Math. **43**(1972), 125–137.

[Va1] M. Valdivia, *Resolution of the identity in certain Banach spaces,* Collect. Math. **39**(1988), 127–140.

[Va2] M. Valdivia, *Projective resolutions of identity in C(K) spaces,* Arch. Math. **54**(1990), 493–498.

[VWZ] J. Vanderwerff, J. H. M. Whitfield and V. Zizler, *Markuševič bases and Corson compacta in duality,* Canadian J. Math. **46**(1994), 200–211.

[V] L. Vašák, *On a generalization of weakly compactly generated Banach spaces,* Studia Math. **70**(1981), 11–19.

[Y] D. Yost, *Asplund spaces for beginners,* Acta Univ. Carolinae **34**(1993), 159–177.

[YS] D. Yost, B. Sims, *Banach spaces with many projections,* Proc. Centre Math. Analysis, Australian National University **14**(1986), 335–342.

[Za] L. Zajíček, *Smallness of sets of nondifferentiability of convex functions,* Czechoslovak Math. J. **41**(1991), 288–296.

[Zi1] N. V. Živkov, *"Generic Gâteaux differentiability of locally Lipschitzian functions,"* Constructive Function Theory 1981, Bulgarian Acad. Sci., Sofia 1983, pp. 590–594.

[Zi2] N. V. Živkov, *Generic Gâteaux differentiability of directionally differentiable mappings,* Rev. Roumaine Math. Pures. Appl. **32**(1987), 179–188.

[Z1] V. Zizler, *Locally uniformly rotund renorming and decomposition of Banach spaces,* Bull. Australian Math. Soc. **29**(1984), 259–265.

[Z2] V. Zizler, *Renorming concerning Mazur's intersection property of balls for weakly compact convex sets,* Math. Annalen **276**(1986), 61–66.

Index

Asplund generated (i.e. GSG) space **16**, 17, 19–23, 26, 28, 29, 31, 32, 35, 46, 51, 58, 63, 65, 67, 76, 78, 93, 133, 139, 144, 154, 155, 157–159, 165, 168

Asplund set **22**, 23, 24, 28, 35, 36, 69, 157, 158

Asplund space **6**, 7–10, 17, 18, 20–24, 26, 28, 29, 32–34, 36, 37, 46, 47, 51, 56, 62, 79–81, 90, 93, 99, 100, 108, 112, 144, 146, 147, 150–155, 159, 168

Baire class one mapping 134, 135, **144**, 145–148, 151, 167

Baire space **2**, 47, 53–55, 57, 58, 60–65, 71, 72, 80

Banach-Mazur game (play) 66, **70**, 71–73, 100, 143

class $\mathcal{A}$ **19**, 20, 21, 29, 35, 51, 63, 65, 93

completely regular space **2**

convex function **4**, 6, 17, 23–25, 65, 80, 90

convex hull **2**

Corson compact 49, 108, **131**, 132, 134, 140–142, 144, **153**, 154–156, 158–161, 167, 168

density **2**

dentable space **33**

Eberlein compact **10**, 12, 13, 26, 30, 31, 35, 76–78, 80, 81, 93, 117, 121, 127, 131, 141, 142, 144, 155, 156, 158, 159, 161, 165, 167, 168

ϵ-tree **33**

fragmentable

– Banach space (class $\widetilde{\mathcal{F}}$) 79, 81, **90**, 91, 93, 94, 96–100, 117, 132, 170

– topological space (class $\mathcal{F}$) **81**, 86, 90, 91, 93, 94, 96, 99, 100, 133, 142, 143, 168, 170

Fréchet derivative **4**, 150

Fréchet differentiability **3**, **4**, 6, 23, 90

Fréchet smooth **4**, 33, 34, 79, 80, 99, 150

Gâteaux derivative **3**

Gâteaux differentiability **3**, 6, 17, 18, 23, 49, 50, 58, 65

Gâteaux differentiability space **37**, 38, 39, 41, 45, 48, 49, 74, 170

Gâteaux smooth **4**, 34, 49, 67, 70, 72–75, 79, 80, 81, 96–99, 168

G_δ point **2**, 37, 38, 42–45, 49, 50

Gul'ko compact 117, **121**, 122-125, 127, 129, 130, 132–136, 141, 142, 144, 155, 161, 167

$\mathcal{K}$-analytic space **67**, **142**

$\mathcal{K}$-countably determined ($\mathcal{K}$-c.d.) space 117, **119**, 120–125, 129, 131, 135, 142

Lindelöf space **120**, 121, 132

locally uniformly rotund (LUR) norm **3**, 33, 34, 100, 116, 133, 139

Markuševič basis 101, **112**, 114, 116, 127, 152

M-differentiability **24**, 25

minimal usco mapping **51**, 52–66, 68, 71, 72, 80, **88**, 90, **146**, 147

CANADIAN MATHEMATICAL SOCIETY SERIES OF MONOGRAPHS AND ADVANCED TEXTS

Monographies et Études de la Société Mathématique du Canada

EDITORS/RÉDACTEURS: Jonathan M. Borwein and Peter B. Borwein

Frank H. Clarke
*Optimization and Nonsmooth Analysis
Erwin Klein and Anthony C. Thompson
**Theory of Correspondences: Including Applications to
Mathematical Economics*
I. Gohberg, P. Lancaster, and L. Rodman
Invariant Subspaces of Matrices with Applications
Jonathan M. Borwein and Peter B. Borwein
*Pi and the AGM—A Study in Analytic Number Theory and
Computational Complexity*
John H. Berglund, Hugo D. Jünghenn, and Paul Milne
*Analysis of Semigroups: Function Spaces Compactifications Representation
Subhashis Nag
The Complex Analytic Theory of Teichmüller Spaces
Manfred Kracht and Erwin Kreyszig
**Methods of Complex Analysis in Partial Differential
Equations with Applications*
Ernest J. Kani and Robert A. Smith
*The Collected Papers of Hans Arnold Heilbronn
Victor P. Snaith
*Topological Methods in Galois Representation Theory
Kalathoor Varadarajan
The Finiteness Obstruction of CTC Wall
F. A. Sherk, P. McMullen, A. Thompson, and A. Weiss
Kaleidoscopes: Selected Writings of H. S. M. Coxeter
Robert V. Moody and Arturo Pianzola
Lie Algebras with Triangular Decompositions
Peter A. Fillmore
A User's Guide to Operator Algebras
Alf van der Poorten
Notes on Fermat's Last Theorem

*Indicates an out-of-print title.
*Indique un titre épuisé.